ADVANCES IN PROSTAGLANDIN AND LEUKOTRIENE RESEARCH

Medical Science Symposia Series

Volume 16

The titles published in this series are listed at the end of this volume.

Advances in Prostaglandin and Leukotriene Research

Basic Science and New Clinical Applications

Edited by

B. Samuelsson
Karolinska Institute, Stockholm, Sweden

R. Paoletti
Department of Pharmacological Sciences, University of Milan, Milan, Italy

G.C. Folco
Department of Pharmacological Sciences, University of Milan, Milan, Italy

E. Granström
Karolinska Hospital, Stockholm, Sweden

and

S. Nicosia
Department of Pharmacological Sciences, University of Milan, Milan, Italy

KLUWER ACADEMIC PUBLISHERS
DORDRECHT / BOSTON / LONDON

Fondazione Giovanni Lorenzini, Medical Science Foundation, Milan, Italy
Giovanni Lorenzini Medical Foundation, Houston, Texas, U.S.A.

A C.I.P. Catalogue record for this book is available from the Library of Congress.

ISBN 1-4020-0146-0

Published by Kluwer Academic Publishers,
P.O. Box 17, 3300 AA Dordrecht, The Netherlands.

Sold and distributed in North, Central and South America
by Kluwer Academic Publishers,
101 Philip Drive, Norwell, MA 02061, U.S.A.

In all other countries, sold and distributed
by Kluwer Academic Publishers,
P.O. Box 322, 3300 AH Dordrecht, The Netherlands.

Printed on acid-free paper

Printed in the Netherlands.

CONTENTS

X CANCER

XI ISOPROSTANES

PREFACE

The *11th International Conference on* ADVANCES IN PROSTAGLANDIN AND LEUKOTRIENE RESEARCH: BASIC SCIENCE AND NEW CLINICAL APPLICATIONS was held, as in the past, in Florence, Italy, on June 4-8, 2000. A selection of lectures presented at the meeting is included in this volume, which continues the tradition of covering both basic and clinical aspects of prostaglandin and leukotriene research.

At this meeting particular emphasis was given to the lipid mediator circuits of lipoxin and leukotrienes and to the expression of the COX2 gene. The plenary lectures by C.N. Serhan (Boston, Massachusetts, U.S.A.) and H.R. Herschman (Los Angeles, California, U.S.A.) were devoted to these areas, while S. Narumiya (Kyoto, Japan) and J.M. Drazen (Boston, Massachusetts, U.S.A.) presented the latest advances in the fields of prostanoid receptors and the genetics of asthma.

Specific sessions focused on vascular physiopathology, cell differentiation and cancer, and the cardiovascular system. Lipoxygenases were discussed in detail at the molecular and cellular levels and attention was given to signal transduction pathways for leukotrienes and prostanoids. K.F. Austen (Boston, Massachusetts, U.S.A.) and T. Izumi (Tokyo, Japan) clarified the role and regulation of LTC_4 synthase of leukotriene biosynthesis. Other important sessions described regulation and control of nitric oxide, cannabimimetic eicosanoids, and the roles of eicosanoids in neuroscience. Applied clinical sessions devoted to the respiratory system and inflammation, including cyclooxygenase pathways, concluded the meeting.

As traditionally has been done in the past meetings of this series, a considerable amount of new material was presented and discussed this year, with particular attention given to the newest clinical data. Due to the unanimous decision of the attendees, this series of meetings will continue to be held.

We would like to express our gratitude to the staff of the Fondazione Giovanni Lorenzini for their important work in connection with the organization of the Conference and to Ann S. Jackson of the Houston office of the Giovanni Lorenzini Medical Foundation for her editorial assistance in the preparation of this volume.

We are also grateful to Dr. Gian Enrico Rovati and Dr. Angelo Sala, the Organizing and Advisory Committees, and the International Program Committee for their valuable contributions.

It is also a pleasure to acknowledge the generous financial support from a number of pharmaceutical companies.

We hope that the rapid publication of this volume, which was made possible by the timely submission of the papers by the authors, will further this exciting area of research.

The Editors

LIST OF CONTRIBUTORS

Cameron Ackerley, *Integrative Biology Programme, The Hospital for Sick Children, 555 University Avenue, Toronto, Ontario M5G 1X8, Canada*

Elisabetta Antonelli, *Dipartimento di Medicina Clinica e Sperimentale, Clinica di Gastroenterologia ed Epatologia; Università degli Studi di Perugia, Perugia, Italy*

K.E. Armour, *Department of Medicine and Therapeutics, University of Aberdeen Medical School, Foresterhill, Aberdeen AB25 2ZD, U.K.*

K.J. Armour, *Department of Medicine and Therapeutics, University of Aberdeen Medical School, Foresterhill, Aberdeen AB25 2ZD, U.K.*

K. Frank Austen, M.D., *Harvard Medical School and Brigham & Women's Hospital, Smith Building, Room 638, One Jimmy Fund Way, Boston, Massachusetts 02115, U.S.A.*

Angelika G. Börsch-Haubold, *Department of Pharmacology, University of Oxford, Mansfield Road, Oxford OX1 3QT, U.K.*

Olivier Boutaud, *Department of Pharmacology, 536 RRB, Vanderbilt University, Nashville, Tennessee 37232-6602, U.S.A.*

Cynthia Brame, *Departments of Pharmacology and Medicine, Vanderbilt University, Nashville, Tennessee 37232-6602, U.S.A.*

Jessica Brannon, *Center for Experimental Therapeutics and Reperfusion Injury, Department of Anesthesiology, Perioperative and Pain Medicine, Brigham and Women's Hospital and Harvard Medical School, 75 Francis Street, Boston, Massachusetts 02115, U.S.A.*

Richard M. Breyer, *Department of Medicine, Vanderbilt University School of Medicine, Nashville, Tennessee 37232-2372, U.S.A.*

Thomas G. Brock, *Division of Pulmonary and Critical Care Medicine, The University of Michigan Health System, Ann Arbor, Michigan 48109-0642, U.S.A.*

Fahdi Chehne, *University of Aleppo, Alsahba Box 16209, Aleppo, Syria*

Yan Chen, *Departments of Pharmacology and Medicine, Vanderbilt University, Nashville, Tennessee 37232-6602, U.S.A.*

Nan Chiang, *Center for Experimental Therapeutics and Reperfusion Injury, Department of Anesthesiology, Perioperative and Pain Medicine, Brigham and Women's Hospital and Harvard Medical School, 75 Francis Street, Boston, Massachusetts 02115, U.S.A.*

Kevin B.S. Chow, *Department of Pharmacology, The Chinese University of Hong Kong, Shatin, Hong Kong SAR, China*

Maria Cirino, *Merck Frosst Center for Therapeutic Research. P.O. Box 1005. Dorval-Pointe-Claire, Quebec, H9R 4P8, Canada*

Clary B. Clish, *Center for Experimental Therapeutics and Reperfusion Injury, Department of Anesthesiology, Perioperative and Pain Medicine, Brigham and Women's Hospital and Harvard Medical School, 75 Francis Street, Boston, Massachusetts 02115, U.S.A.*

Flavio Coceani, *Scuola Superiore S. Anna, Via Carducci 40, 56127 Pisa, Italy*

Sean P. Colgan, *Center for Experimental Therapeutics and Reperfusion Injury, Department of Anesthesiology, Perioperative and Pain Medicine, Brigham and Women's Hospital and Harvard Medical School, 75 Francis Street, Boston, Massachusetts 02115, U.S.A.*

Bruce N.Cronstein, *Department of Medicine, New York University School of Medicine, New York, New York 10024, U.S.A.*

Sean S. Davies, *Departments of Pharmacology and Medicine, Vanderbilt University, Nashville, Tennessee 37232-6602, U.S.A.*

P. Del Soldato, *NiCox SA, Sophia Antipolis, France*

Danielle Denis, *Merck Frosst Center for Therapeutic Research. P.O. Box 1005. Dorval-Pointe-Claire, Quebec, H9R 4P8, Canada*

Michael Dicay, *Mucosal Inflammation Research Group, Faculty of Medicine, University of Calgary, Calgary, Alberta, Canada*

Alain Doglio, *Laboratoire de Virologie, Faculte de Medecine, 06107 Nice, France*

Valentina Fabbroni, *Department of Internal Medicine, University of Florence, Viale G.B. Morgagni 85, 50134 Florence, Italy*

Ornella Fantappiè, *Department of Internal Medicine, University of Florence, Viale G.B. Morgagni 85, 50134 Florence, Italy*

Steven J. Feinmark, *Department of Pharmacology and Center for Molecular Therapeutics, College of Physicians and Surgeons, Columbia University, New York, New York 10032, U.S.A.*

David Fenard, *Laboratoire de Virologie, Faculté de Medecine, 06107 Nice, France*

Stefano Fiorucci, *Dipartimento di Medicina Clinica e Sperimentale, Clinica di*

Gastroenterologia ed Epatologia; Università degli Studi di Perugia, Perugia, Italy

Giancarlo Folco, *Center for Cardiopulmonary Pharmacology, Institute of Pharmacological Sciences, University of Milan, Via Balzaretti 9, 20133 Milan, Italy*

Motonari Fukui, *Department of Respiratory Medicine, Shiga Medical Center for Adults 5-4-30 Moriyama Moriyama, Shiga 524-8524, Japan*

R. Michael Garavito, *522 Biochemistry Building, Michigan State University, East Lansing, Michigan 48824, U.S.A.*

Narinder Gautam, *Department of Physiology and Pharmacology, Karolinska Institute, S-171 77 Stockholm, Sweden*

Michael H. Gelb, *Departments of Chemistry and Biochemistry, University of Washington, Box 351700, Seattle, Washington 98195 U.S.A.*

Mark W. Geraci, *Division of Pulmonary Sciences and Critical Care Medicine, Department of Medicine, University of Colorado Health Sciences Center, 4500 E. Ninth Ave, Denver, Colorado 80262, U.S.A.*

Tracey L. Gesell, *Division of Pulmonary Sciences and Critical Care Medicine, Department of Medicine, University of Colorado Health Sciences Center, 4500 E. Ninth Ave, Denver, Colorado 80262, U.S.A.*

Cristiana Griffoni, *Department of Experimental Biology, University of Bologna, Via Selmi 3, 40126 Bologna, Italy*

Karsten Gronert, *Center for Experimental Therapeutics and Reperfusion Injury, Department of Anesthesiology, Perioperative and Pain Medicine, Brigham and Women's Hospital and Harvard Medical School, 75 Francis Street, Boston, Massachusetts 02115, U.S.A.*

Hiromi Hanaka, *Department of Biochemistry and Molecular Biology, Faculty of Medicine, The University of Tokyo, Hongo 7-3-1, Bunkyo-ku, Tokyo, 113-0033 Japan*

Osamu Hayaishi, *Osaka Bioscience Institute, 6-2-4 Furuedai, Suita City, Osaka 565-0874, Japan*

Tove Hammarberg, *Department of Medical Biochemistry and Biophysics, Division of Physiological Chemistry II, Karolinska Institutet, S-171 77 Stockholm, Sweden*

Hiromi Hanaka, *Department of Biochemistry and Molecular Biology, Faculty of Medicine, The University of Tokyo, Hongo 7-3-1, Bunkyo-ku, Tokyo, 113-0033 Japan*

Per Hedqvist, *Department of Physiology and Pharmacology, Karolinska Institute, S-171 77*

Stockholm, Sweden

Ying Hefner, *Departments of Chemistry and Biochemistry, University of Washington, Box 351700, Seattle, Washington 98195 U.S.A.*

Harvey R. Herschman, *Departments of Biological Chemistry and Molecular and Medical Pharmacology, 341 Molecular Biology Institute, UCLA, 611 Charles E. Young Drive East, Los Angeles, California 90095-1570, U.S.A.*

Tetsuya Hirabayashi, *Department of Biochemistry and Molecular Biology, Faculty of Medicine, The University of Tokyo, Hongo 7-3-1, Bunkyo-ku, Tokyo, 113-0033 Japan*

Stephen Holgate, *Divisions of Respiratory Cell and Molecular Biology, Centre for Infection, Allergy, Inflammation, and Repair, University of Southampton School of Medicine, Southampton, U.K.*

Peter Howarth, *Divisions of Respiratory Cell and Molecular Biology, Centre for Infection, Allergy, Inflammation, and Repair, University of Southampton School of Medicine, Southampton, U.K.*

Atsushi Ichikawa, *Faculty of Pharmaceutical Sciences, Kyoto University, Kyoto 606-8501, Japan*

Yoshiyuki Ide, *Department of Biochemistry and Molecular Biology, Faculty of Medicine, The University of Tokyo, Hongo 7-3-1, Bunkyo-ku, Tokyo, 113-0033 Japan*

Kazunori Ishimura, *Department of Anatomy of Cell Biology, School of Medicine, The University of Tokushima, 3-18-15 Kuramoto, Tokushima 770-8503, Japan*

Seiji Ito, *Department of Medical Chemistry, Kansai Medical University, 10-15 Fumizono, Moriguchi, Osaka 570-8506, Japan*

Takashi Izumi, *Department of Biochemistry and Molecular Biology, Faculty of Medicine, The University of Tokyo, Hongo 7-3-1, Bunkyo-ku, Tokyo 113-0033, Japan and Department of Biochemistry, Faculty of Medicine, The University of Gunma, Showamachi 3-39-22, Maebashi, Gunma, 371-8511 Japan*

Miao Jiang, *Department of Pharmacology, New York Medical College, Valhalla, New York 10595, U.S.A.*

Katsuo Kanmatsuse, *Second Department of Internal Medicine, Nihon University School of Medicine, Ooyaguchi-kamimachi, 30-1, Itabashi-ku, Tokyo 173-8610, Japan*

Kazuhiko Kato, *Pharmaceutical Research Center, Meiji Seika Kaisha, Ltd., Yokohama 222-8567, Japan*

JacSue Kehoe, *Laboratoire de Neurobiologie, Ecole Normale Supérieure, Paris, France*
Robert L. Keith, *Division of Pulmonary Sciences and Critical Care Medicine, Department of Medicine, Denver VA Medical Center, 1055 Clermont St. Box 111A, Denver, Colorado 80220, U.S.A.*

Lois Kelsey, *Integrative Biology Programme, The Hospital for Sick Children, 555 University Avenue, Toronto, Ontario M5G 1X8, Canada*

Jenny Klemm, *Department of Medical Biochemistry and Biophysics, Division of Physiological Chemistry II, Karolinska Institutet, S-171 77 Stockholm, Sweden*

Rao Koduri, *Departments of Chemistry and Biochemistry, University of Washington, Seattle, Washington 98195, U.S.A.*

Karen M. Lakkides, *513 Biochemistry Building, Michigan State University, East Lansing, Michigan 48824, U.S.A.*

Bing K. Lam, *Harvard Medical School and Brigham & Women's Hospital, Smith Building, Room 628, One Jimmy Fund Way, Boston, Massachusetts 02115, U.S.A.*

Michal Laniado Schwartzman, *Department of Pharmacology, New York Medical College, Valhalla, New York 10595, U.S.A.*

Michel Lazdunski, *Institut de Pharmacologie Moléculaire et Cellulaire, UPR 411 CNRS, 660, route des Lucioles, Sophia Antipolis, 06560 Valbonne, France*

Gérard Lambeau, *Institut de Pharmacologie Moléculaire et Cellulaire, UPR 411 CNRS, 660, route des Lucioles, Sophia Antipolis, 06560 Valbonne, France*

Lennart Lindbom, *Department of Physiology and Pharmacology, Karolinska Institute, S-171 77 Stockholm, Sweden*

Graziana Lupattelli, *Institute of Internal Medicine, Angiology and Atherosclerosis, University of Perugia, Italy*

Alvin M. Malkinson, *Department of Pharmaceutical Sciences, University of Colorado Health Sciences Center, 4500 E. Ninth Ave, Denver, Colorado 80262, U.S.A.*

Michael G. Malkowski, *522 Biochemistry Building, Michigan State University, East Lansing, Michigan 48824, U.S.A.*

Jackleen Marji, *Department of Pharmacology, New York Medical College, Valhalla, New York 10595, U.S.A.*

Pier Francesco Mannaioni, *Department of Preclinical and Clinical Pharmacology, University of Florence, Viale G. Pieraccini 6, 50139 Florence, Italy*

Lawrence J. Marnett, *Department of Biochemistry, 850A RRB, Vanderbilt University, School of Medicine, Nashville, Tennessee 37232-6602, U.S.A.*

Roberta Marulli, *Department of Experimental Biology, University of Bologna, Via Selmi 3 , 40126 Bologna, Italy*

Emanuela Masini, *Department of Preclinical and Clinical Pharmacology, University of Florence, Viale G. Pieraccini 6, 50139 Florence, Italy*

Paul Masson, *Merck Frosst Center for Therapeutic Research. P.O. Box 1005. Dorval-Pointe-Claire, Quebec, H9R 4P8, Canada*

Roberto Mazzanti, *Department of Internal Medicine, University of Florence, Viale G.B. Morgagni 85, 50134 Florence, Italy*

Webb McKnight, *Mucosal Inflammation Research Group, Faculty of Medicine, University of Calgary, Calgary, Alberta, Canada*

Kathleen M. Metters, *Merck Frosst Center for Therapeutic Research. P.O. Box 1005. Dorval-Pointe-Claire, Quebec, H9R 4P8, Canada*

Reneé J. Micielli, *513 Biochemistry Building, Michigan State University, East Lansing, Michigan 48824, U.S.A.*

York E. Miller, *Division of Pulmonary Sciences and Critical Care Medicine, Department of Medicine, Denver VA Medical Center, 1055 Clermont St., Denver, Colorado 80220, U.S.A.*

M. Carmen Montesinos, *Department of Medicine, New York University School of Medicine, New York, New York 10024, U.S.A.*

Mark D. Moore, *Division of Pulmonary Sciences and Critical Care Medicine, Department of Medicine, University of Colorado Health Sciences Center, 4500 E. Ninth Ave, Denver, Colorado 80262, U.S.A.*

Ikuo Morita, *Tokyo Medical and Dental University, 1-5-45, Yushima, Bunkyo-ku, Tokyo 113-8549, Japan*

Jason D. Morrow, *Departments of Pharmacology and Medicine, Vanderbilt University, Nashville, Tennessee 37232-6602, U.S.A.*

Sei-itsu Murota, *Tokyo Medical and Dental University, 1-5-45, Yushima, Bunkyo-ku, Tokyo 113-8549, Japan*

Marcelo N. Muscará, *Mucosal Inflammation Research Group, Faculty of Medicine, University of Calgary, Calgary, Alberta, Canada*

Tomohiro Nakayama, *Second Department of Internal Medicine, Nihon University School of Medicine, Ooyaguchi-kamimachi, 30-1, Itabashi-ku, Tokyo 173-8610, Japan*

Francois Nantel, *Merck Frosst Center for Therapeutic Research. P.O. Box 1005. Dorval-Pointe-Claire, Quebec, H9R 4P8, Canada*

Shuh Narumiya, *Department of Pharmacology, Kyoto University Faculty of Medicine, Sakyo, Kyoto 606-8501, Japan*

Santosh Nigam, *Eicosanoid Research Division, Department of Gynecology & Obstetrics, University Medical Centre Benjamin Franklin, Free University Berlin, D-12200 Berlin, Germany*

Mikio Nishizawa, *Department of Medical Chemistry, Kansai Medical University, 10-15 Fumizono, Moriguchi, Osaka 570-8506, Japan*

Denis Normandin, *Merck Frosst Center for Therapeutic Research. P.O. Box 1005. Dorval-Pointe-Claire, Quebec, H9R 4P8, Canada*

John A. Oates, *Department of Pharmacology and Medicine, 536 MRB-I, Vanderbilt University, Nashville, Tennessee 37232-6602, U.S.A.*

Gary P. O'Neill, *Merck Frosst Center for Therapeutic Research. P.O. Box 1005. Dorval-Pointe-Claire, Quebec, H9R 4P8, Canada*

Mitsue Onodera, *Tokyo Medical and Dental University, 1-5-45, Yushima, Bunkyo-ku, Tokyo 113-8549, Japan*

Bengt Persson, *Department of Medical Biochemistry and Biophysics, Division of Physiological Chemistry I, Karolinska Institutet, S-171 77 Stockholm, Sweden*

Marc Peters-Golden, M.D., *Division of Pulmonary and Critical Care Medicine, The University of Michigan Health System, Ann Arbor, Michigan 48109-0642, U.S.A.*

Michael Pillinger,, *Department of Medicine, New York University School of Medicine, New York, New York 10024, U.S.A.*

Patrick Provost, *Department of Medical Biochemistry and Biophysics, Division of Physiological Chemistry II, Karolinska Institutet, S-171 77 Stockholm, Sweden*

Olof Rådmark, *Department of Medical Biochemistry and Biophysics, Division of Physiological Chemistry II, Karolinska Institutet, S-171 77 Stockholm, Sweden*

Laura Raimondi, *Department of Preclinical and Clinical Pharmacology, University of Florence, Viale G. Pieraccini 6, 50139 Florence, Italy*

Stuart H. Ralston, *Department of Medicine and Therapeutics, University of Aberdeen Medical School, Polwarth Building, Foresterhill, Aberdeen AB25 2ZD, U.K.*

Konda Veera Reddy, *Department of Medical Biochemistry and Biophysics, Division of Physiological Chemistry II, Karolinska Institutet, S-171 77 Stockholm, Sweden*

Srinivasa T. Reddy, *Departments of Biological Chemistry and Molecular and Medical Pharmacology, 341 Molecular Biology Institute, UCLA, 611 Charles E. Young Drive East, Los Angeles, California 90095-1570, U.S.A.*

Duolikun. Rehemudula, *Second Department of Internal Medicine, Nihon University School of Medicine, Ooyaguchi-kamimachi, 30-1, Itabashi-ku, Tokyo 173-8610, Japan*

Massimo Riccio, *Institute of Cytomorphology, C.N.R., Bologna, Italy*

C. Jill Rieke, *513 Biochemistry Building, Michigan State University, East Lansing, Michigan 48824, U.S.A.*

L. Jackson Roberts, II, *Departments of Pharmacology and Medicine, Vanderbilt University, Nashville, Tennessee 37232-6602, U.S.A.*

Marek Rola-Pleszczynski, *Immunology Division, Department of Pediatrics, Faculty of Medicine, Université de Sherbrooke, Sherbrooke QC J1H 5N4, Canada*

Anthony Sampson, *Divisions of Respiratory Cell and Molecular Biology, Centre for Infection, Allergy, Inflammation, and Repair, University of Southampton School of Medicine, Southampton, U.K.*

Bengt Samuelsson, *Department of Medical Biochemistry and Biophysics, Division of Physiological Chemistry II, Karolinska Institutet, S-171 77 Stockholm, Sweden*

Spartaco Santi, *Institute of Cytomorphology, C.N.R., Bologna, Italy*

Iacopo Sardi, *Department of Preclinical and Clinical Pharmacology, University of Florence, Viale G. Pieraccini 6, 50139 Florence, Italy*

Ian Sayers, *Division of Human Genetics, Centre for Infection, Allergy, Inflammation, and Repair, University of Southampton School of Medicine, Southampton, U.K.*

Tankred Schewe, *Eicosanoid Research Division, Department of Gynecology & Obstetrics, University Medical Centre Benjamin Franklin, Free University Berlin, D-12200 Berlin, Germany*

Dietlind Schumann, *Institute of Pharmaceutical Chemistry, University of Frankfurt, Marie-Curie-Straße 9, 60439 Frankfurt, Germany*

James H. Schwartz, *Center for Neurobiology and Behavior, College of Physicians and Surgeons, Columbia University, New York, New York 10032, U.S.A.*

Eric Seidlitz, *Integrative Biology Programme, The Hospital for Sick Children, 555 University Avenue, Toronto, Ontario M5G 1X8, Canada*

Charles N. Serhan, *Center for Experimental Therapeutics and Reperfusion Injury, Department of Anesthesiology, Perioperative and Pain Medicine, Brigham and Women's Hospital and Harvard Medical School, 75 Francis Street, Boston, Massachusetts 02115, U.S.A.*

Pattabhiraman Shankaranarayanan, *Eicosanoid Research Division, Department of Gynecology & Obstetrics, University Medical Centre Benjamin Franklin, Free University Berlin, D-12200 Berlin, Germany*

Takao Shimizu, *Department of Biochemistry and Molecular Biology, Faculty of Medicine, The University of Tokyo, Hongo 7-3-1, Bunkyo-ku, Tokyo 113-0033, Japan*

Helmut Sinzinger, *Department of Nuclear Medicine, University of Vienna, Austria, Währinger Gürtel 18-20, A-1090 Vienna, Austria*

William L. Smith, *513 Biochemistry Building, Michigan State University, East Lansing, Michigan 48824, U.S.A.*

Masayoshi Soma, *Second Department of Internal Medicine, Nihon University School of Medicine, Ooyaguchi-kamimachi, 30-1, Itabashi-ku, Tokyo 173-8610, Japan*

Enzo Spisni, *Department of Experimental Biology, University of Bologna, Via Selmi 3, 40126 Bologna, Italy*

Jana Stankova, *Immunology Division, Department of Pediatrics, Faculty of Medicine, Université de Sherbrooke, Sherbrooke QC J1H 5N4, Canada*

Dieter Steinhilber, *Institute of Pharmaceutical Chemistry, University of Frankfurt, Marie-Curie-Straße 9, 60439 Frankfurt, Germany*

Yukihiko Sugimoto, *Faculty of Pharmaceutical Sciences, Kyoto University, Kyoto 606-8501, Japan*

Ying Sun, *513 Biochemistry Building, Michigan State University, East Lansing, Michigan 48824, U.S.A.*

Mark Sutherland, *Eicosanoid Research Division, Department of Gynecology & Obstetrics, University Medical Centre Benjamin Franklin, Free University Berlin, D-12200 Berlin, Germany*

Toshiko Suzuki-Yamamoto, *Department of Anatomy of Cell Biology, School of Medicine, The University of Tokushima, 3-18-15 Kuramoto, Tokushima 770-8503, Japan*

Dagmar Szellas, *Institute of Pharmaceutical Chemistry, University of Frankfurt, Marie-Curie-Straße 9, 60439 Frankfurt, Germany*

Kan Terawaki, *Department of Biochemistry and Molecular Biology, Faculty of Medicine, The University of Tokyo, Hongo 7-3-1, Bunkyo-ku, Tokyo 113-0033, Japan*

Maryse Thivierge, *Immunology Division, Department of Pediatrics, Faculty of Medicine, Université de Sherbrooke, Sherbrooke QC J1H 5N4, Canada*

Elizabeth D. Thuresson, *513 Biochemistry Building, Michigan State University, East Lansing, Michigan 48824, U.S.A.*

Tamara L. Tieman, *Department of Pharmacology, College of Physicians and Surgeons, Columbia University, New York, New York 10032, U.S.A.*

Vittorio Tomasi, *Department of Experimental Biology, University of Bologna, Via Selmi 3, 40126 Bologna, Italy*

Yoshihiro Urade, *Department of Molecular Behavioral Biology, Osaka Bioscience Institute, 6-2-4 Furuedai, Suita City, Osaka 565-0874, Japan, Tel: 81-6-6872-4851, Fax: 81-6-6872-2841*

Emmanuel Valentin, *Institut de Pharmacologie Moléculaire et Cellulaire, UPR 411 CNRS, 660, route des Lucioles, Sophia Antipolis, 06560 Valbonne, France*

R.J. van 't Hof, *Department of Medicine and Therapeutics, University of Aberdeen Medical School, Foresterhill, Aberdeen AB25 2ZD, U.K.*

David J. Wadleigh, *Departments of Biological Chemistry and Molecular and Medical Pharmacology, 341 Molecular Biology Institute, UCLA, 611 Charles E. Young Drive East, Los Angeles, California 90095-1570, U.S.A.*

Helmar Waiczies, *Eicosanoid Research Division, Department of Gynecology & Obstetrics, University Medical Centre Benjamin Franklin, Free University Berlin, D-12200 Berlin, Germany*

J.L. Wallace, *Department of Pharmacology and Therapeutics, University of Calgary, 3330*

Hospital Drive NW, Calgary, Alberta, T2N 4N1, Canada
Jining Wang, *Department of Biochemistry, The University of Tokushima School of Medicine, Tokushima 770-8503, Japan*

Mong-Heng Wang, *Department of Pharmacology, New York Medical College, Valhalla, New York 10595, U.S.A.*

Kikuko Watanabe, *Division of Applied Life Sciences, Graduate School of Integrated Sciences and Arts, University of East Asia, 2-1 Ichinomiyagakuencho, Shimonoseki, Yamaguchi 751-8503 Japan*

Steve P. Watson, *Department of Pharmacology, University of Oxford, Mansfield Road, Oxford OX1 3QT, U.K.*

Gerald Weissmann, *Department of Medicine, New York University School of Medicine, New York, New York 10024, U.S.A.*

Oliver Werz, *Institute of Pharmaceutical Chemistry, University of Frankfurt, Marie-Curie-Straße 9, 60439 Frankfurt, Germany*

Byron Wingerd, *513 Biochemistry Building, Michigan State University, East Lansing, Michigan 48824, U.S.A.*

Helen Wise, *Department of Pharmacology, The Chinese University of Hong Kong, Shatin, Hong Kong SAR, China*

Simon Wong, *Merck Frosst Center for Therapeutic Research. P.O. Box 1005. Dorval-Pointe-Claire, Quebec, H9R 4P8, Canada*

Kei Yamamoto, *Department of Biochemistry, The University of Tokushima School of Medicine, Tokushima 770-8503, Japan*

Shozo Yamamoto, *Department of Biochemistry, The University of Tokushima School of Medicine, Tokushima 770-8503, Japan*

Takehiko Yokomizo, *Department of Biochemistry and Molecular Biology, Faculty of Medicine, The University of Tokyo, Hongo 7-3-1, Bunkyo-ku, Tokyo 113-0033, Japan*

Ji-liang Zhao, *Division of Rheumatology, Immunology, and Allergy,* M.D., *Harvard Medical School and Brigham & Women's Hospital, Smith Building, Room 638, One Jimmy Fund Way, Boston, Massachusetts 02115, U.S.A.*

Phosphorylation and Activation of 5-Lipoxygenase by P38 Kinase-Activated MAPKAP Kinases

Oliver Werz, Jenny Klemm, Olof Rådmark, and Bengt Samuelsson

5-lipoxygenase (5-LO) catalyzes the formation of leukotriene A_4 from arachidonic acid. Factors determining 5-LO enzyme activity are calcium, phosphatidylcholine (membranes), ATP, and the cellular redox status. Also, a possible role of phosporylation events for 5-LO activity in the cell has been discussed. In particular, connection to the activity of protein kinase C in alveolar macrophages [1] and the effects of protein tyrosine kinase inhibitors in HL-60 cells and polymorphonuclear leukocytes (PMNL) [2] indicated such a role. The three MAP kinase families, the extracellular regulated kinases (ERKs), the c-jun N-terminal kinases/stress activated protein kinases (JNKs/SAPKs), and the p38 MAP kinases, have been implicated in a variety of cellular functions, including cell proliferation, differentiation, and immune responses [3, 4]. The ERKs are activated mainly by mitogens such as growth factors and G-protein coupled receptor agonists, while JNKs and p38 are activated by various types of cellular stress.

For three different cell types (polymorphonuclear leukocytes, the monocytic cell line Mono Mac 6 (MM6), and the B-lymphocyte cell line BL41-E95-A) we found that upregulation of p38-activated MAPKAP kinases, which could phosphorylate 5-LO *in vitro*, occurred under conditions which also upregulated 5-LO activity of these cells. Some of the results are described in detail in reference [5].

Mono Mac 6 cells

By in-gel kinase assay, which gives an estimate of the molecular weights of kinases which can phosphorylate a particular substrate *in vitro*, samples from MM6 cells were found to contain 5-LO kinase activities. In Figure 1 it is illustrated that samples from ionophore-activated cells gave phosphorylated bands at approximately 40, 47, 55, 80, and 100 kDa, in the 5-LO containing gel. The bands at 80 and 100 kDa were present regardless of ionophore stimulation of the cells, and also without addition of 5-LO to the gels, indicating that these were due to phosphorylation of kinases in the cell sample. However, the bands at 40, 47, and 55 kDa were prominent only after stimulation of the cells, and when 5-LO was present in the gel. MM6 cells were pretreated with three different kinase inhibitors for 30 minutes prior to stimulation with A23187. Cell lysates were prepared and subjected to in-gel kinase assay. The MEK inhibitor U0126 (prevents the activation of ERK1 and 2) and

B. Samuelsson et al. (eds.), Advances in Prostaglandin and Leukotriene Research, 1–7.

the protein kinase C inhibitor GF109203x had only slight effects. However, the specific p38 inhibitor SB203580 potently reduced the phosphorylated bands at 40, 47, and at 55 kDa. Thus, p38 was involved in the activation of 5-LO kinases.

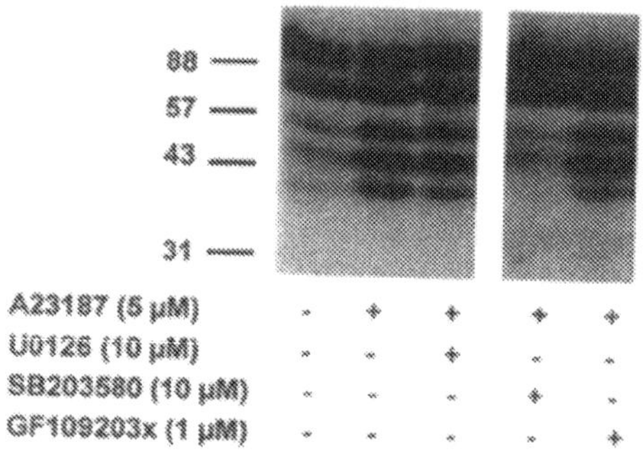

Figure 1. Effects of kinase inhibitors on phosphorylation of 5-LO, determined by in-gel kinase assay. Differentiated MM6 cells (2.5 x 10^7 in 1 ml PBS with 1 mg/ml glucose and 1 mM $CaCl_2$) were preincubated for 30 minutes at 37°C with inhibitors. Cells were then stimulated with ionophore (5 μM). After 3 minutes incubations were terminated by addition of SDS loading buffer, and heating at 95°C for 6 minutes. Aliquots of total cell lysates (corr. 0.25 x 10^6 cells) were electrophoresed on a 10% SDS-PAGE gel containing 0.15 mg/ml purified recombinant human 5-LO. After electrophoresis, gels were washed to remove SDS. Proteins in the gel were denatured by incubation for 1 hour at RT in 50 mM Tris-HCl, pH 8, 20 mM DTT, 2 mM EDTA, 6 M guanidine-HCl. To renature proteins, gels were washed overnight in 50 mM Tris-HCl pH 8, 1 mM DTT, 2 mM EDTA, 0.04% Tween 20. After preincubation at RT for 1 hour in 30 ml kinase buffer (20 mM HEPES pH 7.6, 20 mM $MgCl_2$, 25 mM β-glycerophosphate, 10 mM 4-nitrophenylphosphate, 2 mM DTT, 0.2 mM Na_3VO_4), gels were finally incubated in 10 ml of kinase buffer containing 50 μM ATP and 10 μCi/ml [γ-^{32}P]ATP, for 1 hour at 30°C with shaking. To remove unreacted [γ-^{32}P]ATP, gels were washed extensively (2 days) followed by drying (*in vacuo*) and autoradiography.

MAPKAP kinase 2 (MK2) phosphorylates its substrates at serine within the consensus motif hyd-Xaa-Arg-Xaa-Xaa-Ser, and this motif is present in 5-LO (Ser-271). Since MK2 is phosphorylated and activated by the upstream p38 kinase, it was reasonable that MK2 is the kinase in MM6 cells which phosphorylated 5-LO. Indeed, MK2 immunoprecipitates and purified MK2 from rabbit skeletal muscle were active with 5-LO as substrate. Also, the MWs of the kinase activities at 47 and 55 kDa were compatible with MK2. The kinase activity migrating at 40 kDa probably corresponds to MK3, another MAPKAP kinase activated by p38.

MM6 cells incubated with only ionophore A23187 gave low 5-LO activity. However, priming of the cells with phorbol myristate actetate (PMA) gave 7-fold increased activity (Figure 2). This upregulation of activity was accompanied by translocation to nuclear membranes and increased capacity for phosphorylation of 5-LO *in vitro*. These responses occurred within 3-5 minutes, and they were counteracted by the protein kinase inhibitor calphostin C. Also in presence of exogenous arachidonate, PMA-priming gave a 1.5- to 2-fold upregulation of 5-LO activity (not shown). These results agree with the concept that 5-LO should translocate to the nuclear membrane, where 5-lipoxygenase activating protein is located, and where endogenous arachidonate is released (by $cPLA_2$), for efficient LTA_4 production. Also, it appears that phosphorylation events is/are involved.

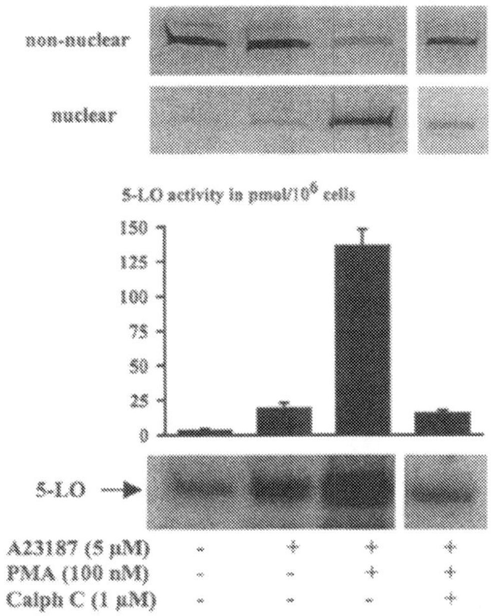

Figure 2. PMA primes MM6 cells for activation of 5-LO. MM6 cells in PBS were preincubated with calphostin C for 30 minutes, and primed with PMA for 10 minutes, as indicated. Then, cells were stimulated with A23187 (5 µM). To determine subcellular localization of 5-LO, 1 x 10^7 cells were lyzed by 0.1% NP-40, and nuclear and non-nuclear fractions (800 x g pellet and supernatant) were analyzed for 5-LO by Western blot. To determine 5-LO activity, metabolites formed in 10 minutes incubations (3 x 10^6 cells in 1 ml) were analyzed by HPLC (mean ± S.E., n=3). To determine phosphorylation of 5-LO, incubations (2.5 x 10^7 cells in 1 ml) were stopped by addition of ice-cold stop-buffer, MK2 was immunoprecipitated, and kinase activity of the MK2-IPs towards 5-LO was determined by *in vitro* kinase assay. The IP was mixed with purified 5-LO in kinase buffer (25 mM HEPES pH 7.5, 25 mM $MgCl_2$, 25 mM β-glycerophosphate, 2 mM DTT, 0.1 mM Na_3VO_4) containing ATP (100 µM) and [γ-^{32}P]ATP (2 µCi/ml). Final volume was 20 µl, and incubation time 30 minutes at 30°C. The reaction was terminated by SDS loading buffer and heating at 95°C for 6 minutes. Samples were separated by SDS-PAGE, and phosphorylated proteins were visualized by autoradiography of the dried gel.

Polymorphonuclear Leukocytes (PMNL)

Also when samples from PMNL were subjected to in-gel kinase assay, the three bands at 55, 47, and 40 kDa appeared (Figure 3), compatible with phosphorylation of 5-LO by MK2 and MK3. When the cells were preincubated with the p38 kinase inhibitor SB203580, these bands were practically absent. Different types of stimuli could activate MK2 and MK3 in PMNL, as A23187, thapsigargin, fMLP, PMA, TNFα, and the well-known p38 activator sodium arsenite (SA, which induces cell stress).

Compared to ionophore, platelet activating factor (PAF) is a less efficient activator of leukotriene synthesis. We also found quite low 5-LO activity in PAF-stimulated PMNL (Figure 3 right panel). Also, cell stimulation with PAF (with or without arachidonate) did not result in a detectable increase in 5-LO kinase activity (Figure 3, left panel). However, when PAF and sodium arsenite were combined (together with exogenous arachidonic acid) 5-LO activity increased about 4-fold, coinciding with strong 5-LO kinase activation, comparable to that obtained with A23187. Without addition of arachidonate, sodium arsenite and PAF did not give increased 5-LO activity. Also, sodium arsenite did not upregulate the already high activity obtained in ionophore stimulated cells (not shown). Thus, it appears that activation of 5-LO kinases could upregulate 5-LO activity in PMNL, when induced by an agent (PAF) which itself gave poor stimulation of 5-LO kinases.

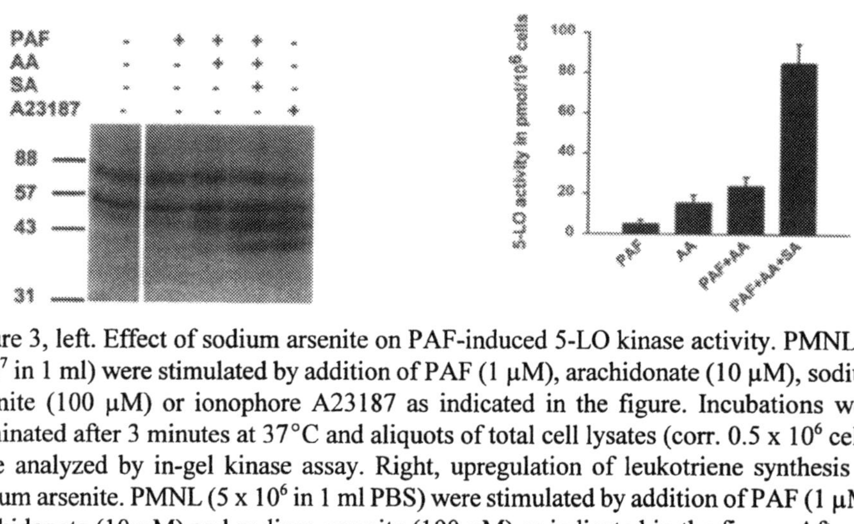

Figure 3, left. Effect of sodium arsenite on PAF-induced 5-LO kinase activity. PMNL (5 x 10^7 in 1 ml) were stimulated by addition of PAF (1 μM), arachidonate (10 μM), sodium arsenite (100 μM) or ionophore A23187 as indicated in the figure. Incubations were terminated after 3 minutes at 37°C and aliquots of total cell lysates (corr. 0.5 x 10^6 cells) were analyzed by in-gel kinase assay. Right, upregulation of leukotriene synthesis by sodium arsenite. PMNL (5 x 10^6 in 1 ml PBS) were stimulated by addition of PAF (1 μM), arachidonate (10 μM) and sodium arsenite (100 μM) as indicated in the figure. After 10 minutes at 37°C 5-LO activity was determined. Results represent the mean ± S.E., n=3.

B-lymphocyte Cell Line BL41-E95-A

Leukotriene synthesis in BL41-E95-A cells was determined under stress conditions. As shown in Figure 4, osmotic shock with sorbitol (0.4 to 1 M) upregulated leukotriene synthesis (induced with ionophore plus arachidonate) up to 5-fold compared to control incubations which received only ionophore plus arachidonate. Treatment of BL41-E95-A cells with sodium arsenite (SA, 0.1 to 1 mM), which mimics heat shock, gave a similar increase in leukotriene formation. Finally, the inflammatory cytokines TNFα and IL-1 (added together with A23187 plus arachidonate), also enhanced cellular leukotriene synthesis. These different stress stimuli are known to activate p38, and we also confirmed activation of MK2, leading to increased phosphorylation of the MK2 substrate heat shock protein 27 (not shown). Also, increased phosphorylation of 5-LO *in vitro* was found by in-gel kinase assay of lysates from BL41-E95-A cells.

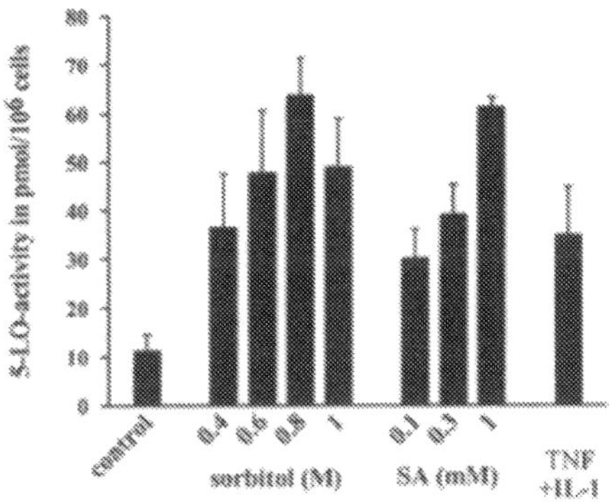

Figure 4. Cell stress stimulates 5-LO activity in BL41-E95-A cells. Cells (1.5 x 10^7 in 1 ml PBS) were incubated for 10 minutes at 37°C with the indicated stimuli, and 5-LO activity was determined by HPLC. All incubations received 10 μM ionophore A23187 plus 40 μM AA. Sorbitol, sodium arsenite, or TNFα and IL-1 (1 ng/ml each) were added as indicated.

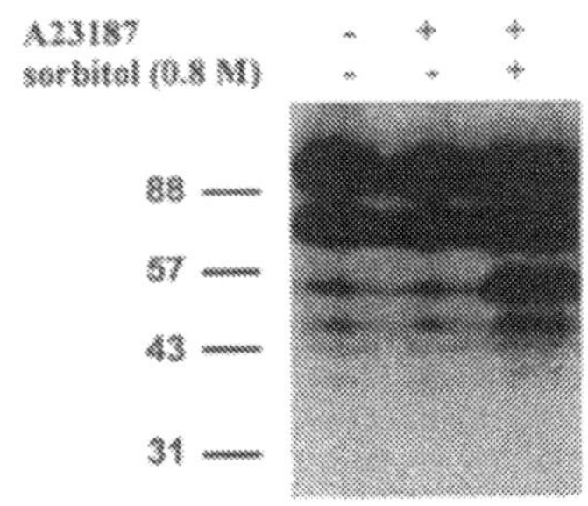

Figure 5. Osmotic stress upregulates 5-LO kinase activity in BL41-E95-A cells. Cells (2.5 x 10^7 in 1 ml PBS containing 1mg/ml glucose and 1 mM $CaCl_2$) were incubated at 37°C with the indicated additives. After 3 minutes, cells were lysed by addition of SDS loading buffer, vortexed and heated at 95°C for 6 min. Aliquots corresponding to 0.5 x 10^6 cells were analyzed for 5-LO kinase by in-gel kinase assay.

In contrast to MM6 cells, stimulation of BL41-E95-A with ionophore alone did not give enhanced 5-LO kinase activity, however simultaneous addition of sorbitol led to activation of 5-LO kinases at 47 and 55 kD (Figure 5). Migration properties and sensitivity aginst SB203580, indicated that 5-LO kinases also in samples from BL41-E95-A cells, is p38-regulated MK2. The band at 55 kDa was rather strong and less sensitive to the p38 inhibitor, apparently this band was due also to other kinase activities. When MM6 cells were treated with both ionophore and sorbitol, no further kinase activation was observed.

The p38 inhibitor SB203580 was used to confirm the involvement of p38 kinase in stress-induced leukotriene synthesis. BL41-E95-A cells were preincubated with SB203580, stimulated with NaCl (together with ionophore and arachidonate) and 5-LO activity was determined. As shown in Figure 6, 5-LO activity in intact BL41-E95-A cells was dose-dependently inhibited by SB203580, with IC50 3-5 μM. In broken cell preparations, the compound did not affect crude 5-LO activity up to 30 μM, and the IC50 value for purified recombinant 5-LO was > 100 μM (data not shown). This suggests involvement of p38 in the signal transduction pathway leading to stress-induced leukotriene synthesis in BL41-E95-A cells.

In view of these observations for three different cell types, it appears probable that phosphorylation of Ser-271, within the MK2 motif present in 5-LO, is one of the factors determining the cellular activity of 5-LO.

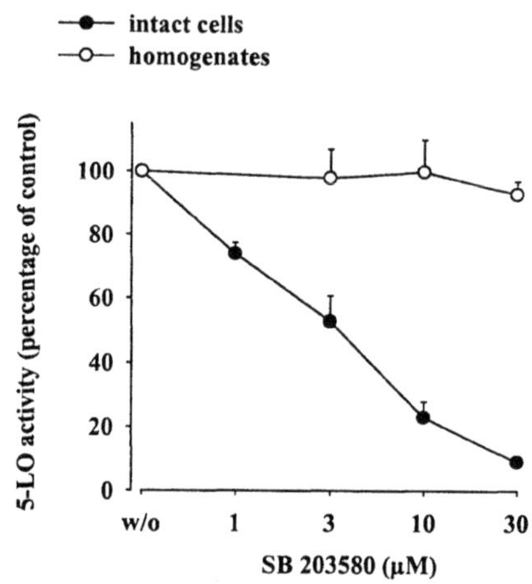

Figure 6. Effect of SB203580 on 5-LO activity in BL41-E95-A cells. Intact cells (1.5×10^7 in PBS containing 1mg/ml glucose and 1 mM $CaCl_2$) or corresponding homogenates were preincubated with the indicated concentrations of SB203580 for 30 minutes at 37°C (intact cells) or 4°C (homogenates). To intact cells NaCl (0.3 M) was added together with ionophore A23187 and arachidonate (10 μM and 40 μM, respectively). To homogenates 1 mM ATP was added and the 5-LO reaction was started by addition of $CaCl_2$ and arachidonate (1 mM and 40 μM, respectively). After 10 minutes at 37°C, 5-LO activity was determined by HPLC. Results are expressed as mean ± S.E. of three independent experiments.

Acknowledgment

We thank Agneta Nordberg for excellent technical assistance. This study was supported by grants from the Swedish Medical Research Council (03X-217), from the European Union (BMH4-CT96-0229), and from the Verum foundation. O.W. received a Karolinska Institute Guest Scientist fellowship.

References

1. Peters-Golden M, McNish RW, Sporn PH, Balazovich K. Basal activation of protein kinase C in rat alveolar macrophages: implications for arachidonate metabolism. Am J Physiol 1991;261:L462-71.
2. Lepley RA, Muskardin DT, Fitzpatrick FA. Tyrosine kinase activity modulates catalysis and translocation of cellular 5-lipoxygenase. J Biol Chem. 1996;271:6179-84.
3. Cano E, Mahadevan LC. Parallel signal processing among mammalian MAPKs. Trends Biochem Sci 1995;20:117-22.
4. Waskiewicz AJ, Cooper JA. Mitogen and stress response pathways: MAP kinase cascades and phosphatase regulation in mammals and yeast. Curr Opin Cell Biol 1995;7:798-805.
5. Werz O, Klemm J, Samuelsson B, Radmark O. 5-Lipoxygenase is phosphorylated by p38 kinase dependent MAPKAP kinases. Proc Natl Acad Sci USA. 2000;97:5261-66.

5-LIPOXYGENASE AND THE NUCLEUS: WHERE, WHEN, HOW, AND WHY?

Marc Peters-Golden and Thomas G. Brock

Only in the last decade has research focused on the intracellular site(s) of leukotriene (LT) synthesis. These studies have revealed an unexpected critical role for the cell nucleus as a site at which 5-lipoxygenase (5-LO) and other LT biosynthetic enzymes are localized. This chapter will review the current understanding of the compartmentalization of LT biosynthesis. The localization of 5-LO in resting and activated leukocytes under different physiologic and pathophysiologic conditions will be discussed, with emphasis on the molecular mechanisms and catalytic and noncatalytic implications of enzyme compartmentalization.

The Recognition of a LT Biosynthetic Metabolon at the Nuclear Envelope

The first decade of research into LT biosynthesis was dominated by efforts to characterize the relevant enzymes, their cofactors, and their cellular distribution. Little was known about the intracellular sites at which these enzymes functioned. A foundation for our current understanding of such matters derives from studies conducted in the late 1980s and early 1990s, in which crude fractionation methods were used to study the three key proteins involved in the initiation of LT synthesis: cytosolic phospholipase A_2 (cPLA_2), 5-LO, and 5-LO activating protein (FLAP). This work demonstrated that, in both resting and activated cells, FLAP was confined to a particulate subcellular fraction, presumably a membrane site [1]. By contrast, cPLA_2 [2] and 5-LO [3] were found in the soluble compartment of resting leukocytes, but were redistributed in a Ca^{2+}-dependent fashion to a particulate fraction upon activation.

It was inferred from these early studies that all three proteins were co-localized at the same cellular membrane. Moreover, because LTs were recognized to be secreted from leukocytes and to act as paracrine mediators, it was widely assumed that the membrane site at which they were assembled was the plasma membrane. However, none of these assumptions was justified on the basis of the crude methods employed in these initial studies.

In the early to mid-1990s, this issue was definitively addressed by several laboratories in a series of studies that utilized more comprehensive subcellular fractionation methods as well as confocal and ultrastructural immunomicroscopic techniques. Indeed, this body of work demonstrated co-localization of all three proteins in activated leukocytes of various types, including macrophages, monocytes, neutrophils, and mast cells. Surprisingly, however, the site at which these proteins were co-localized was not the plasma membrane, but the nuclear envelope. FLAP, an arachidonic acid-binding protein

B. Samuelsson et al. (eds.), Advances in Prostaglandin and Leukotriene Research, 9–15.

[4], was constitutively localized to the nuclear envelope and perinuclear endoplasmic reticulum [5,6]. $cPLA_2$ was found to translocate from its resting locale in the cytosol to the nuclear envelope upon activation [6-9], and most arachidonic acid released by the actions of $cPLA_2$ was hydrolyzed from nuclear membrane phospholipids [9]. 5-LO was likewise found to translocate to the nuclear envelope upon activation [5,6,10-13]. It was subsequently discovered that LTC_4 synthase, like FLAP, is an integral nuclear envelope protein [14]. Therefore, all the enzymatic steps involved in LTC_4 synthesis were localized to this membrane site.

A macromolecular complex of sequential enzymes in a metabolic pathway has been termed a "metabolon" [15]. Such a complex can be assumed to result in a cellular microenvironment that allows efficient channeling of metabolic intermediates. There may well be circumstances in which $cPLA_2$ or 5-LO translocate to and function at intracellular sites other than the nuclear envelope, but it would be expected that deviations from the above paradigm would compromise the metabolic efficiency of LT biosynthesis. An important implication of the metabolon concept is that the site of macromolecular assembly is not a random event, but has evolved in a manner that serves the "wisdom" of the cell. Given that LTs are known to be secreted extracellularly, it is entirely counter-intuitive that their synthesis occurs at the nuclear envelope. This fact strongly suggests that they must subserve important functions within or near the cell nucleus.

The Complexity of 5-LO Compartmentalization in Resting Leukocytes

Further surprises regarding the localization of 5-LO, in particular, were yet to come. First, the locale of 5-LO in resting leukocytes was found to be cell-specific. The protein was exclusively cytosolic in resting peritoneal macrophages [6], as well as monocytes [10], neutrophils [16], and eosinophils [17,18] isolated from peripheral blood. By contrast, mast cells [12,16] and alveolar macrophages [10,11] contained both cytosolic and intranuclear pools of 5-LO. A constant feature in all cell types was that LT biosynthesis was associated with the translocation of 5-LO to the nuclear envelope. Of course, its site of origin in the cytosol or the nucleoplasm of a resting cell would dictate its eventual association with either the outer or the inner nuclear membrane, respectively, upon activation.

The next surprise was that, even within a given cell type, 5-LO was found to be capable of shuttling in or out of the nucleus in response to *in vivo* or experimental conditions. Such movement is unassociated with enzyme activation and LT synthesis, and is to be distinguished from the Ca^{2+}-dependent association with nuclear envelope that accompanies LT synthesis. This shuttling between compartments is presumed to occur via the nuclear pore complex. Examples of dynamic regulation of 5-LO compartmentalization will be discussed later.

Molecular Mechanisms Mediating Compartmentalization of 5-LO

Although we have learned a great deal about where 5-LO is localized under different circumstances, our knowledge of the molecular mechanisms which dictate localization is quite rudimentary. It is highly likely that enzyme compartmentalization is determined directly by interactions between specific molecular motifs within the protein and structural components of the cell, or indirectly by interactions with intermediate partner proteins. Candidate motifs within 5-LO that may mediate Ca^{2+}-dependent translocation to the nuclear envelope include putative Ca^{2+}-binding and lipid-binding domains in the N-terminal β-barrel region [19], a src homology-3 binding domain near the C-terminus that may interact with cytoskeletal proteins [20], and one or more of the many consensus sites for phosphorylation by various protein kinases, including tyrosine kinases [21]. By contrast, shuttling of 5-LO through the nuclear pores of resting cells is almost certainly mediated by motifs functioning as nuclear import and export sequences that bind to chaperones capable of interacting with the nuclear pore complex. Evidence supports a role for a bipartite nuclear import sequence rich in basic residues near the enzyme's C-terminus [22], and several leucine-rich motifs distributed throughout the protein might serve as candidate nuclear export sequences [23]. It is also likely that the activities of such nuclear import and export sequences are further regulated by phosphorylation reactions.

Dynamic Regulation of 5-LO Compartmentalization

The compartmentalization of 5-LO is subject to dynamic modulation by such factors as culture conditions [17,24], cellular recruitment to tissues [17,24,25], and cytokine exposure [18]. Two examples of this dynamic regulation that have been well studied are discussed for illustrative purposes. The first is observed in the course of tissue-specific macrophage differentiation. Monocytes circulating in peripheral blood are the precursors for all tissue macrophages. They emigrate from the bloodstream, home to various tissues, and differentiate in a tissue-specific fashion into mature tissue macrophages. Monocytes themselves contain predominantly cytosolic 5-LO [10]. Interestingly, this cytosolic distribution is preserved during the course of monocyte differentiation into peritoneal macrophages [6] as well as pulmonary interstitial macrophages [25]. However, the final stage of differentiation into a pulmonary alveolar macrophage, i.e. the process of macrophage emigration through the alveolar wall and into the alveolar space, is uniquely associated with import of 5-LO into the nucleoplasm. This process appears to occur relatively quickly upon cellular entry into the alveolar space, as a prominent intranuclear pool of enzyme is found in all alveolar macrophages, regardless of their duration of residence in the airspace [25]. This distribution of 5-LO in alveolar macrophages, apparently unique among mononuclear phagocytes, may be a consequence of cellular exposure to the unique alveolar milieu. This conclusion is supported by the fact that 5-LO distribution gradually reverts back to the monocyte/non-alveolar macrophage pattern (i.e. predominantly cytosolic pool) during removal of these cells from the alveolar compartment and culture *ex vivo* over a period of three days [25]. This reversal of 5-LO distribution pattern could reflect either export of 5-LO molecules formerly found in the nucleoplasm, or turnover of the intranuclear 5-LO and a failure to import into the nucleus the newly synthesized polypeptide molecules. In preliminary experiments, the nuclear export inhibitor leptomycin failed to prevent this redistribution over time, suggesting a failure of import in cells removed from the alveolus. On the basis of these observations, we suggest that factors unique to the alveolar milieu favor nuclear accumulation of 5-LO, primarily by promoting nuclear import. The precise factors and molecular mechanisms responsible for such import remain to be defined.

A second example of dynamic regulation of 5-LO compartmentalization can be observed in eosinophils. Eosinophils isolated from peripheral blood and studied in suspension contain a predominant cytosolic pool of 5-LO. Upon adherence of these cells on fibronectin, import of 5-LO into the nucleus can be observed over a 30-60 minute interval [17]. Such an intranuclear pool of enzyme is also seen in eosinophils that adhere to the vascular wall and emigrate from the bloodstream in the process of recruitment to tissues [17].

Effects of 5-LO Compartmentalization on LT Synthetic Capacity

How do these shifts in intracellular distribution influence the capacity for LT synthesis upon subsequent activation? Interestingly, nuclear import of 5-LO increases LT synthetic capacity in fully differentiated alveolar macrophages, but reduces it in adherent eosinophils. These disparate consequences of nuclear import in the two cell types can be attributed, at least in part, to cell-specific differences in the capacities of different enzyme pools to translocate to the nuclear envelope in response to agonist stimulation. Thus, the intranuclear pool of 5-LO in alveolar macrophages translocates readily to the nuclear envelope, while the cytosolic pool does not [11]. In contrast, the intranuclear pool of enzyme in adherent eosinophils translocates far less well upon stimulation than does the cytosolic pool in eosinophils studied in suspension [17]. These results reinforce the critical importance of enzyme translocation for 5-LO activation. A further contrast to both of these situations is provided by the mast cell, in which both cytosolic and intranuclear pools of 5-LO readily translocate to the nuclear envelope upon agonist activation [11]. The reasons for cell- and compartment-specific differences in enzyme translocatability are not known at present, but likely reflect differences in the distribution of critical protein kinases and/or of 5-LO partners which act as anchors or chaperones. It should be acknowledged that contrary results have been obtained with eosinophils under different experimental circumstances. In particular, incubation of eosinophils for 6 hours with interleukin-5 has been shown to increase intranuclear 5-LO as well as LT synthetic capacity upon agonist stimulation; however, translocation of enzyme following stimulation was not examined [18].

Another general means by which 5-LO localization may influence metabolic function of the enzyme involves its capacity to interact with downstream enzymes. The major 5-LO product of alveolar macrophages is LTB_4. Synthesis of this molecule requires the hydrolysis of 5-LO-derived LTA_4 by the enzyme LTA_4 hydrolase. Preliminary data derived from both *in situ* (immunohistochemical staining of lung tissue) and *ex vivo* (immunofluorescence microscopy and subcellular fractionation of isolated cells) studies indicate that in alveolar macrophages, LTA_4 hydrolase is found in both cytosolic and intranuclear compartments. In contrast, this enzyme appears to be largely cytosolic in monocytes as well as in structural lung cells such as epithelial cells. These data suggest that nuclear import of the hydrolase accompanies that of 5-LO which occurs during macrophage differentiation in the alveolar space. The nuclear import of both enzymes in parallel may facilitate efficient coupling of these two sequential reactions at the nucleoplasmic face of the inner nuclear membrane of alveolar macrophages.

Finally, one can envision situations in which compartmentalization of this pivotal enzyme is altered in association with disease states. This could occur on the basis of either inherited or acquired alterations in any of the cellular components that mediate its intracellular distribution, including molecular motifs within 5-LO itself, in signal transduction cascades, or in partner proteins. Increases or decreases in LT biosynthetic capacity could then occur as a result of such alterations, contributing to disease expression.

Consequences of the Nuclear Localization of 5-LO

It is now well accepted that, under most circumstances, translocation of 5-LO to the nuclear envelope is a critical component of enzyme activation. This process allows the enzyme to 1) gain access to substrate arachidonic acid which is hydrolyzed from nuclear membrane phospholipids and presented by FLAP and 2) couple with downstream enzymes necessary for LT synthesis. The surprising observation that the LT biosynthetic metabolon is found at the nuclear envelope, rather than other membrane sites, surely suggests that LTs or alternative metabolic byproducts such as reactive oxygen species possess important autocrine actions in or near the nucleus. An important goal of future research will be to identify the nuclear targets for the actions of 5-LO products. There is precedent for nuclear eicosanoid receptors [26-28], and the types and roles of such receptors mediating direct intranuclear actions of 5-LO metabolites remain to be elucidated.

In addition, it is becoming increasingly apparent that cells have evolved elaborate mechanisms to be able to dynamically regulate the distribution of 5-LO between cytosol and nucleoplasm of resting cells. There is evidence that this represents one means for modulating the output of the LT biosynthetic pathway. Future studies must elucidate how compartmentalization influences the ability of 5-LO to translocate upon activation and to couple with upstream and downstream components of the metabolon. It is also appealing to speculate that 5-LO might subserve functions beyond its traditional catalytic role. The rationale for this possibility derives from the fact that the enzyme is capable of interacting with other molecules [20,29], and from the observations that even in resting cells not synthesizing LTs, it has been identified in both the euchromatin region of the nucleus (where active gene transcription occurs) [10] and tightly associated with chromatin [16]. The latter findings suggest the intriguing possibility that 5-LO might regulate transcriptional events in the nucleus.

References

1. Miller DK, Gillard JW, Vickers PJ, et. al. Identification and isolation of a membrane protein necessary for leukotriene production. Nature 1990;343(6255):278-81.
2. Channon J, Leslie C. A calcium-dependent mechanism for associating a soluble arachidonoyl-hydrolyzing phospholipase A_2 with membrane in the macrophage cell line RAW 264.7. J Biol Chem 1990;265:5409-13.
3. Rouzer CA, Kargman S. Translocation of 5-lipoxygenase to the membrane in human leukocytes challenged with ionophore A23187. J Biol Chem 1988;263(22):10980-88.
4. Mancini J, Abramovitz M, Cox M, et. al. 5-lipoxygenase-activating protein is an arachidonate binding protein. FEBS Lett 1993;318:277-81.
5. Woods J, Evans J, Ethier D, et. al. 5-Lipoxygenase and 5-lipoxygenase activating protein are localized in the nuclear envelope of activated human leukocytes. J Exp Med 1993;178:1935-46.
6. Peters-Golden M, McNish R. Redistribution of 5-lipoxygenase and cytosolic phospholipase A_2 to the nuclear fraction upon macrophage activation. Biochem Biophys Res Commun 1993;196:147-53.
7. Glover S, Bayburt T, Jonas M, Chi E, Gelb M. Translocation of the 85-kDa phospholipase A_2 from cytosol to the nuclear envelope in rat basophilic leukemia cells stimulated with calcium ionophore or IgE/antigen. J Biol Chem 1995;270:15359-67.
8. Schievella A, Regier M, Smith W, Lin L. Calcium-mediated translocation of cytosolic phospholipase A_2 to the nuclear envelope and endoplasmic reticulum. J Biol Chem 1995;

270:30749-54.
9. Peters-Golden M, Song K, Marshall T, Brock T. Translocation of cytosolic phospholipase A_2 to the nuclear envelope elicits topographically localized phospholipid hydrolysis. Biochem J 1996;318:797-803.
10. Woods J, Coffey M, Brock T, Singer I, Peters-Golden M. 5-Lipoxygenase is located in the euchromatin of the nucleus in resting human alveolar macrophages and translocates to the nuclear envelope upon cell activation. J Clin Invest 1995;95:2035-40.
11. Brock TG, McNish RW, Peters-Golden M. Translocation and leukotriene synthetic capacity of nuclear 5-lipoxygenase in rat basophilic leukemia cells and alveolar macrophages. J Biol Chem 1995;270:21652-58.
12. Chen X-S, Naumann T, Kurre U, Jenkins N, Copeland N, Funk C. cDNA cloning, expression, mutagenesis, intracellular localization, and gene chromosomal assignment of mouse 5-lipoxygenase. J Biol Chem 1995;270:17993-99.
13. Pouliot M, McDonald P, Krump E, et. al. Colocalization of cytosolic phospholipase A_2, 5-lipoxygenase, and 5-lipoxygenase-activating protein at the nuclear membrane of A23187-stimulated human neutrophils. Eur J Biochem 1996;238:250-58.
14. Penrose J, Spector J, Lam B, et. al. Purification of human lung LTC_4 synthase and preparation of a polyclonal antibody. Am J Respir Crit Care Med 1995;152:283-89.
15. Srere P. Complexes of sequential metabolic enzymes. Annu Rev Biochem 1987;56:89-124.
16. Brock TG, Paine R, Peters-Golden M. Localization of 5-lipoxygenase to the nucleus of unstimulated rat basophilic leukemia cells. J Biol Chem 1994;269:22059-66.
17. Brock T, Anderson J, Fries F, Peters-Golden M, Sporn P. Decreased LTC_4 synthesis accompanies adherence-dependent nuclear import of 5-lipoxygenase in human blood eosinophils. J Immunol 1999; in press.
18. Cowburn A, Holgate S, Sampson A. IL-5 increases expression of 5-lipoxygenase-activating protein and translocates 5-lipoxygenase to the nucleus in human blood eosinophils. J Immunol 1999;163:456-65.
19. Gillmor S, Villasenor A, Fletterick R, Sigal E, Browner M. The structure of mammalian 15-lipoxygenase reveals similarity to the lipases and the determinants of substrate specificity. Nat Struct Biol 1997;4:1003-9.
20. Lepley RA, Fitzpatrick F. 5-lipoxygenase contains a functional Src homology 3-binding motif that interacts with the Src homology 3 domain of Grb2 and cytoskeletal proteins. J Biol Chem 1994;269:24163-68.
21. Lepley R, Muskardin D, Fitzpatrick F. Tyrosine kinase activity modulates catalysis and translocation of cellular 5-lipoxygenase. J Biol Chem 1996;271:6179-84.
22. Healy A, Peters-Golden M, Yao J, Brock T. Identification of a bipartite nuclear localization sequence necessary for nuclear import of 5-lipoxygenase. J Biol Chem 1999; 274:29812-18.
23. Nakielny S, Dreyfuss G. Nuclear export of proteins and RNAs. Curr Opin Cell Biol 1997; 9:420-29.
24. Brock T, McNish R, Bailie M, Peters-Golden M. Rapid import of cytosolic 5-lipoxygenase into the nucleus of neutrophils after in vivo recruitment and in vitro adherence. J Biol Chem 1997;272:8276-80.
25. Covin R, Brock T, Bailie M, Peters-Golden M. Altered expression and localization of 5-lipoxygenase accompany macrophage differentiation in the lung. Am J Physiol (Lung Cell Mol Physiol) 1998; 275:L303-L310.
26. Devchand P, Keller H, Peters J, Vazquez M, Gonzalez F, Wahli W. The PPARα-leukotriene B_4 pathway to inflammation control. Nature 1996; 384:39-43.
27. Bhattacharya M, Peri K, Ribeiro-da-Silva A, et. al. Localization of functional prostaglandin E_2 receptors EP_3 and EP_4 in the nuclear envelope. J Biol Chem 1999;274:

15719-24.
28. Bhattacharya M, Peri K, Almazan G, et. al. Nuclear localization of prostaglandin E_2 receptors. Proc Natl Acad Sci USA 1998;95:15792-97.
29. Provost P, Samuelsson B, Radmark O. Interaction of 5-lipoxygenase with cellular proteins. Proc Natl Acad Sci USA 1999;96:1881-85.

Calcium Binding Site of 5-Lipoxygenase

Tove Hammarberg, Patrick Provost, Konda Veera Reddy, Bengt Persson, and Olof Rådmark

5-lipoxygenase (5LO) catalyzes the formation of leukotriene A_4 from arachidonic acid. Factors determining 5LO enzyme activity are calcium, phosphatidylcholine (membranes), ATP, the cellular redox status, and phosphorylation events. Ionophores (increase intracellular Ca^{2+}) are effective, and Ca^{2+} also activates 5LO *in vitro*. 5LO binds Ca^{2+} in a reversible manner with a K_d close to 6 μM and maximum binding around two Ca^{2+} per 5LO [1]. While examining the Ca^{2+} binding site on 5LO, we noticed similarity between the 5LO N-terminal domain, and calcium binding C2-domains of for example $cPLA_2$. Here, we describe that the N-terminal domain of 5LO binds calcium and mediates calcium stimulation of enzyme activity.

The N-terminal sequence of 5LO was modeled into the structure of rabbit reticulocyte 15LO [2] (1LOX) using the program ICM (version 2.7, Molsoft LLC). In support of the model (Figure 1) residues not conserved between 5LO and 15LO, as well as gaps, are found mainly in the loops connecting the eight β-strands. The 5LO model resembled reported structures of C2 domains, for example in $cPLA_2$ (1RLW) [3]. However, it did not exhibit any of the two distinct topologies that have been described for C2 domains [4].

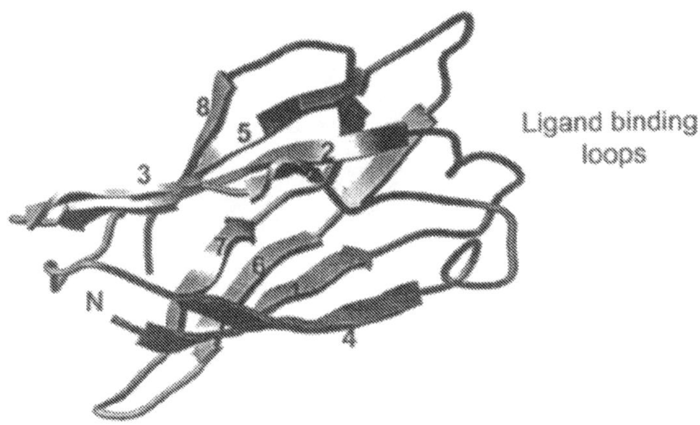

Figure. 1. Model of the N-terminal domain of 5LO.

B. Samuelsson et al. (eds.), Advances in Prostaglandin and Leukotriene Research, 17–20.

Calcium binding to different parts of 5LO was determined. Native 5LO as well as its two putative domains: the N-terminal β-barrel (0-128) and the C-terminal domain (120-673), were expressed as GST fusion proteins in *E. coli*. All three fusion proteins were recovered in insoluble inclusion bodies. $^{45}Ca^{2+}$-overlay was performed directly on the solubilized inclusion bodies without further protein purification, as described [1]. After SDS-PAGE and electrotransfer onto PVDF membrane, the membrane was washed 3 x 10 minutes with 5 mM imidazole buffer (pH 7.4) containing 1 mg/ml octaethylene glycol dodecyl ether ($C_{12}E_8$) and then incubated for 30 minutes in 15 ml of a solution containing 5 mM imidazole buffer (pH 7.4), 1 mg/ml $C_{12}E_8$, 60 mM KCl, 5 mM $MgSO_4$, and 10 μM $^{45}CaCl_2$ (specific radioactivity 1 mCi/μmol). The membrane was washed [30% (v/v) ethanol in de-ionized water] and dried before exposure to Fuji RX film at -70°C for 35-50 hours. The overlay film in Figure 2 shows (i) GST-5LO gave the same response as recombinant 5LO without a GST-tag (lanes 4 and 5); (ii) GST alone did not bind Ca^{2+} (lane 3); and (iii) the N-terminal part (GST-5LO(1-129) binds Ca^{2+} as strongly as the entire GST-5LO. The C-terminal part (GST-5LO(121-673) gave a weak response.

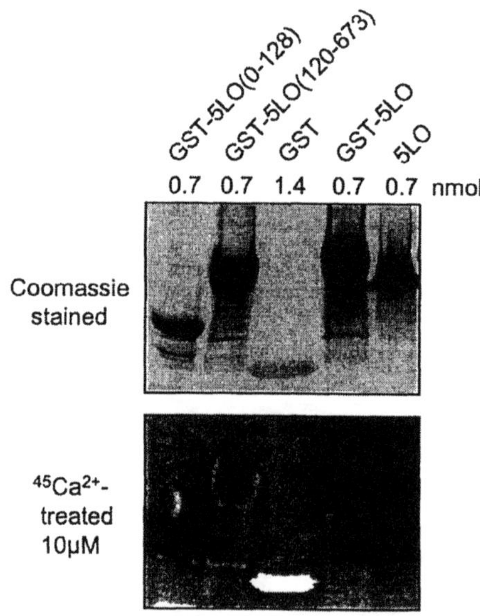

Figure 2. Calcium binding of truncated 5LO proteins fused to GST. After electrophoresis of the same set of samples on two SDS-polyacrylamide gels (4-15%), proteins were electrotransferred onto PVDF membranes. One membrane was Coomassie stained (upper panel) and the other membrane was subjected to $^{45}Ca^{2+}$-overlay (10μM Ca^{2+}) followed by autoradiography (lower panel). The amounts of proteins loaded are indicated.

In $cPLA_2$ and other C2 domain proteins, the Ca^{2+}-chelating ligands are located in the loops connecting the β-sheets. Totally nine putative Ca^{2+}-chelating amino acids, in putative ligand binding loops of 5LO, were mutated to Ala. After expression in *E. coli*, the loop 2 mutant (N43A+D44A+E46A) showed a clearly reduced Ca^{2+} response, but also mutations in the other loops shifted the response slightly towards higher Ca^{2+} concentrations. In determinations of enzyme activity over time, both initial rate and final amount of product were reduced for the loop 2 mutant as compared to control, for example at 10 μM calcium (Figure 3). However, at a high calcium concentration (1 mM) the loop 2 mutant gave a time course that was similar in character to the time course obtained for wild-type 5LO at 10 μM calcium. This connects the calcium affinity of the 5LO β-barrel to calcium stimulation of enzyme activity. We recently reported that Mg^{2+}, in the millimolar concentration range, could stimulate 5LO [5]. Also the Mg^{2+}-stimulation of the loop 2 mutant was drastically impaired (Figure 3) supporting the concept that Mg^{2+} and Ca^{2+} bind to the same site on 5LO.

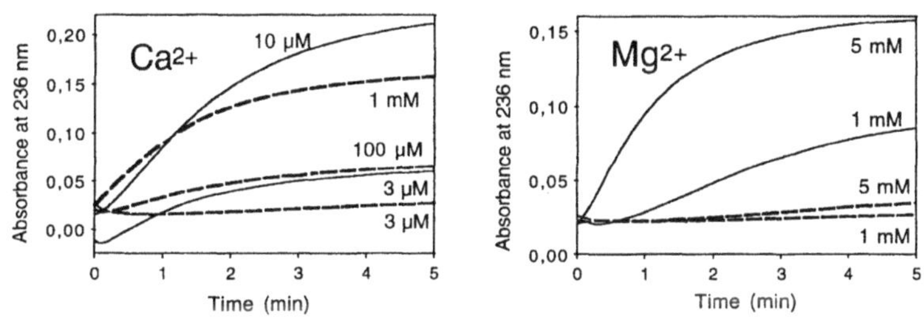

Figure 3. Stimulation of 5LO enzyme activity by Ca^{2+} and Mg^{2+}, time courses for purified wild-type 5LO (intact lines) and loop 2 mutant (broken lines). Purified protein (250 ng) was added to a cuvette containing 500 μl Chelex-treated 50 mM Tris-HCl pH 7.5, 20 μM arachidonate, 25 μg/ml phosphatidylcholine, 2.5 μM 13-HPODE, 1 mM DTT, 1 μM EDTA, 1 μM EGTA, and indicated concentrations of $CaCl_2$ (left) or $MgCl_2$ (right).

In summary, the apparent structural and functional similarities between 5LO and C2 domain proteins ($cPLA_2$) indicate that an N-terminal β-barrel of 5-lipoxygenase functions as a C2 domain, for calcium regulation of enzyme activity.

Acknowledgments

This study was supported by grants from the Swedish Medical Research Council (03X-217, 13X-12564), the European Union (BMH4-CT96-0229, Bio4-CT97-2123), the Swedish Foundation for Strategic Research, the Åke Wiberg Foundation, and the Verum Foundation. PP received fellowships from the Heart and Stroke Foundation of Canada and the Medical Research Council of Canada.

References

1. Hammarberg T, Rådmark O. 5-lipoxygenase binds calcium. Biochemistry 1999;38:4441-47.
2. Gillmor SA, Villasenor A, Fletterick R, Sigal E, Browner MF. The structure of mammalian 15-lipoxygenase reveals similarity to the lipases and the determinants of substrate specificity. Nat Struct Biol 1997;4:1003-9.
3. Perisic O, Fong S, Lynch DE, Bycroft M, Williams RL. Crystal structure of a calcium-phospholipid binding domain from cytosolic phospholipase A_2. J Biol Chem 1998;273: 1596-1604.
4. Nalefski EA, Falke JJ. The C2 domain calcium-binding motif: structural and functional diversity. Protein Sci 1996;5:2375-90.
5. Reddy KV, Hammarberg T, Rådmark O. Mg^{2+} activates 5-lipoxygenase in vitro: dependency on concentrations of phosphatidylcholine and arachidonic acid. Biochemistry 2000;39:1840-48.

Regulation of Cellular 5-Lipoxygenase Activity by Glutathione Peroxidases

Dietlind Schumann, Oliver Werz, Dagmar Szellas, and Dieter Steinhilber

B-lymphocytes, monocytes, and partially differentiated human leukemic cell lines express 5-lipoxygenase protein but no concomitant cellular activity. Recently, we and others could show that glutathione peroxidases (GPx) are responsible for inhibition of 5-lipoxygenase activity in these cell lines [1,2]. Characterization of the endogenous 5-lipoxygenase inhibitors in the B-lymphocytic cell line BL41-E95A and in HL-60 cells demonstrated that the respective inhibitors display properties identical to GPx-4 [2]. Characterization of the functionality of this inhibitor revealed that inhibition of 5-lipoxygenase in these cell lines was abolished when 5-LO activity was determined in cell homogenates in the absence of thiols. In contrast to these findings, monocytic cells contain an endogenous inhibitor which suppresses 5-lipoxygenase activity in intact cells and in cell homogenates both in the presence and absence of thiols (Figure 1). Purification of this endogenous 5-LO inhibitor of monocytic cells followed by protein sequencing and immunoblotting revealed that the inhibitor is GPx-1. In contrast to peroxidase activity, there was no requirement for millimolar concentrations of thiols for 5-LO inhibition indicating that different catalytic properties of peroxidases are addressed by these two actions.

Results and Discussion

5-LO inhibition by selenium-dependent peroxidases was investigated in three different cell lines. HL-60 cells, differentiated for 4 days by 1.5 % DMSO, Mono Mac 6 cells grown for two days in the presence of TGFβ (1ng/ml), and BL41-E95A cells were used as models for granulocytic, monocytic, and B-lymphocytic cells, respectively. The cells were grown either in medium containing 10% fetal calf serum (FCS-medium), under serum-free conditions (SFM) or in serum-free medium supplemented with selenium (Se^{4+}, 1 ng/ml). As can be seen from Figure 1, transfer of HL-60 cells, BL41-E95A, or Mono Mac 6 cells from FCS-medium into serum-free medium for several passages leads to a strong upregulation of cellular 5-LO activity which can be suppressed by addition of selenium to the serum-free medium.

B. Samuelsson et al. (eds.), Advances in Prostaglandin and Leukotriene Research, 21–23.

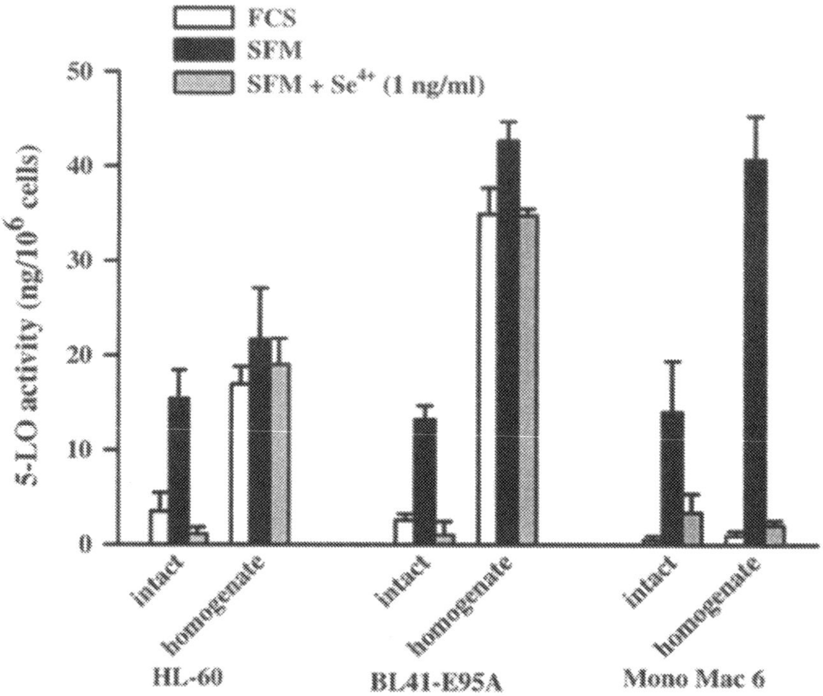

Figure 1. Selenium-dependent inhibition of 5-LO activity of intact cells and cell homogenates from HL-60, BL41-E95A and Mono Mac 6 cells. Cells were cultured under the indicated culture conditions. Then, 5-LO activity of intact cells and cell homogenates was determined in the absence of exogenously added thiols.

However, when 5-LO activity was determined in the corresponding cell homogenates in the absence of exogenously added thiols, there was a similar 5-LO activity under all cell culture conditions in HL-60 cells and in BL41-E95A cells, respectively. In contrast, in cell homogenates of Mono Mac 6 cells, 5-LO activity was low when the cells were cultured in the presence of selenium whereas high 5-LO activity was only observed under selenium free conditions. The data suggest that Mono Mac 6 cells contain a selenium-dependent endogenous 5-LO inhibitor that, in contrast to the inhibitor of B-cells and HL-60 cells which was shown previously to be GPx-4, does not depend on millimolar concentrations of thiols. Purification, protein sequencing, and immunoblotting revealed that the inhibitor is GPx-1. Further analysis of the inhibitory effect demonstrated that GPx-1 does not require millimolar concentrations of thiols for 5-LO inhibition and that there is no direct binding of GPx-1 to 5-LO [3]. As shown in Figure 2, cellular GPx activity of Mono Mac 6 cells is much higher than in HL-60 cells and BL41-E95A cells which could be the reason that GPx-1 is responsible for 5-LO inhibition in Mono Mac 6 cells but not in the other cell lines.

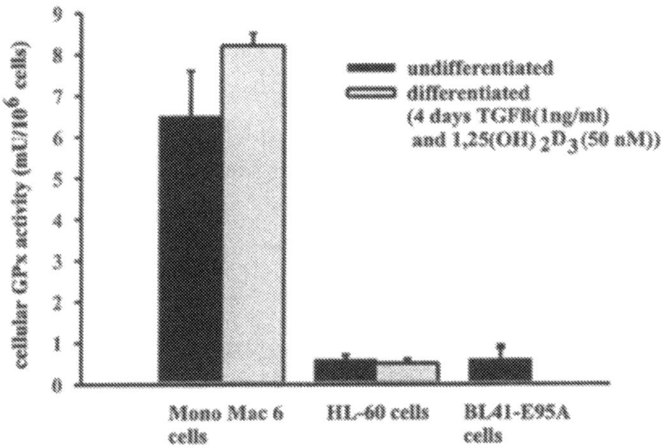

Figure 2. Determination of cellular glutathione peroxidase activity. Cells were grown under the indicated culture conditions ($1{,}25(OH)_2D_3$ = calcitriol). Then, the cells were harvested and cellular glutathione peroxidase activity was determined as described [4].

Interestingly, cellular peroxidase activity does not significantly change during differentiation of the Mono Mac 6 and HL-60 cells (Figure 2). However, in Mono Mac 6 cells there is a 200- to 400-fold increase in cellular activity but only an about 3-fold increase in 5-LO protein expression by calcitriol when the cells were cultured before for 2 days in the presence of TGFβ [5]. These data suggest that cell maturation is accompanied by a strong decrease in 5-LO sensitivity against peroxidases. This process might be involved in the upregulation of cellular 5-LO activity during myeloid cell maturation.

References

1. Weitzel F, Wendel A. Selenoenzymes regulate the activity of leukocyte 5- lipoxygenase via the peroxide tone. J Biol Chem 1993;268:6288-92.
2. Werz O, Steinhilber D. Selenium-dependent peroxidases suppress 5-lipoxygenase activity in B-lymphocytes and immature myeloid cells - the presence of peroxidase-insensitive 5-lipoxygenase activity in differentiated myeloid cells. Eur J Biochem 1996;242:90-97.
3. Straif D, Werz O, Kellner R, Bahr U, Steinhilber D. Glutathione peroxidase-1 but not 4 is involved in the regulation of cellular 5-lipoxygenase activity in monocytic cells. Biochem J 2000;349:455-61.
4. Wendel A. Glutathione peroxidase. Meth Enzymol 1981;77:325-33.
5. Brungs M, Rådmark O, Samuelsson B, Steinhilber D. Sequential induction of 5-lipoxygenase gene expression and activity in Mono Mac 6 cells by transforming growth factor-beta and 1,25-dihydroxyvitamin D3. Proc Natl Acad Sci USA 1995;92:107-11.

Novel Aspects Related to Biosynthesis and Biological Actions of Hepoxilins: Interrelationship with Phospholipid Hydroperoxide Glutathione Peroxidase (PHGPx)

Pattabhiraman Shankaranarayanan, Mark Sutherland, Helmar Waiczies, Tankred Schewe, and Santosh Nigam

Introduction

The 12-lipoxygenase (12-LOX) pathway of AA metabolism in platelets and other cells is bifurcated at the level of the oxygenated product 12-hydroperoxyeicosatetranoic acid (12-HPETE) into a reduction route leading to 12-HETE as a primary product and an isomerization route forming hepoxilins (HXA_3 and HXB_3) as major products. Both 12-HETE and hepoxilins are biologically active eicosanoids and possess a myriad of biological actions [1,2 and references therein]. Glutathione peroxidases, which are key players in the defense of tissues and cells against the oxidative damage [3], are also intimately involved in the regulation of AA metabolism [4-7]. This role has been attributed to glutathione peroxidases for following reasons: (a) they reduce the hydroperoxyfatty acids formed by lipoxygenases or cyclooxygenases to corresponding alcohols, (b) they regulate the hydroperoxide tone of the system, which is contributed largely by the hydroperoxyfatty acids. It is known that a low hydroperoxide tone functions as a stimulus for lipoxygenase and cyclooxygenase reactions but a higher hydroperoxide tone causes suicide inactivation of dioxygenases and of substrates for hydroperoxidase reaction [8]. Four different glutathione peroxidases, which are characterized by the selenocysteine group at the active site, are known up until now [3]: cytosolic glutathione peroxidase (GPx-1), PHGPx (GPx-4), plasma GPx (pGPx), and gastrointestinal GPx (GIGPx). But, only GPx-1 and PHGPx play a significant role in the intracellular metabolism of eicosanoids. Several studies in the past dealing with the role of glutathione peroxidases in the eicosanoid metabolism of platelets considered predominantly GPX-1 as the sole enzyme. To our knowledge no evidence for the coeval occurrence of PHGPx in platelets has been demonstrated so far. The present study provides an evidence for the presence of PHGPx in platelets. Its pivotal role in the regulation of 12-LOX pathway has been demonstrated by the inactivation of glutathione peroxidases by iodoacetate without any inactivation of 12-LOX [9] as well as in selenium-deficient cells.

B. Samuelsson et al. (eds.), Advances in Prostaglandin and Leukotriene Research, 25–29.

Results and Discussion

Western blot analyses with rabbit PHGPx antibody show a strong immunoreactive band of 23 kDa molecular weight in lysates of human platelets and A431 cells (positive control), but not in erythrocyte cytosol, which is known as a rich source for GPx-1 but is deprived of PHGPx (Figure 1). In order to outflank any nonspecificity of the rabbit antibody, a monoclonal antibody raised against human recombinant PHGPx was used to confirm the detection of PHGPx in the platelet cytosol (Figure 1B). Thus, we provided for the first time an evidence for the occurrence of PHGPx protein in human platelets. However, the RT-PCR analysis of PHGPx-mRNA and 12-LOX-mRNA in the lysate of platelets demonstrated a positive reaction for the 12-LOX messenger, but not for the PHGPx messenger (Figure 2), which led us to believe the shorter life-time of the PHGPx messenger as compared to other mRNA species in platelets. This assumption was strongly supported by the RT-PCR analysis of megakaryocytes prepared from the human bone marrow, in which large quantities of both PHGPx and 12-LOX messengers were found (Figure 2). Additional support came from the analysis of the breakdown rate of the PHGPx messenger in the human megakaryoblast cell line UT7 before and after differentiation to megakaryocytes with phorbolester, which depicted a half-life of ~ 3 hours in nondifferentiated cells as compared with only ~ 0.5 hours in differentiated cells. This observation explained clearly why the PHGPx messenger is hardly detectable in platelets. In order to evaluate the distinct roles of PHGPx and GPx-1 in platelets, we prepared selenium-deficient UT7 cells [10], since it is known that the GPx-1, but not PHGPx, is sensitive to selenium deficiency. By replenishing these cells with selenium we obtained cells showing almost no GPx-1 activity but possessing substantial PHGPx activity. Inasmuch as UT7 cells do not contain 12-LOX, the reduction of 12-HPETE to 12-HETE by these cells gave a clear-cut evidence that PHGPx could substitute GPx-1 under specific conditions (data not shown).

Application of iodoacetate, which does not inhibit 12-LOX activity, as an irreversible inhibitor of selenocysteine-containing glutathione peroxidases for studying the AA metabolism in platelets caused a significant shift in the formation of 12-HPETE instead of 12-HETE, whereas the sum of the 2 products was negligibly reduced (Figure 3). A similar pattern of both 12-LOX metabolites was also observed when 12-HPETE was added exogenously (data not shown). Since PHGPx and GPx-1 lower the hydroperoxide tone in the cells, there occurs inevitably a modulation of the AA metabolism. Thus, the formation of hepoxilin is significantly augmented in iodoacetate-treated platelets (0.4 nmols HXA_3 with iodoacetate versus < 0.1 nmol without iodoacetate), whereas almost no hepoxilin was detected in platelets in the absence of iodoacetate. To our knowledge this role of PHGPx has not been reported so far. Considering that hepoxilins are involved in a number of biological effects in mammalian cells [2,11,12], it is reasonable to assume that the shift of the reduction route of 12-LOX pathway to the isomerization route occurs inevitably under oxidative stress situation in order to keep the functional integrity of the platelets. In summary, the PHGPx apparently constitutes an integral part in the fine tuning network of the regulation of 12-LOX pathway of AA metabolism in human platelets.

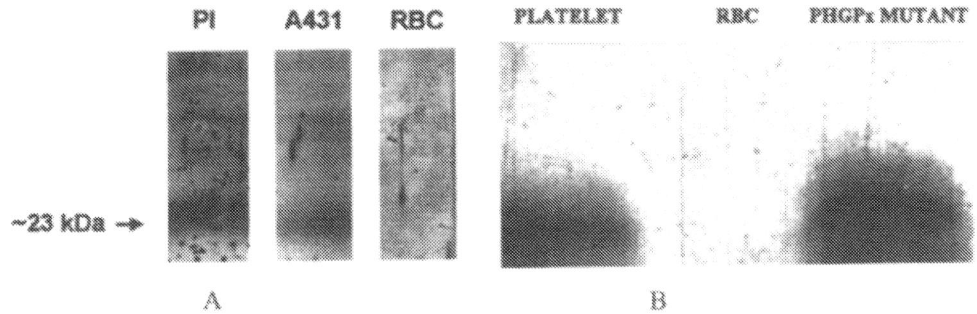

Figure 1. Western blots with two different antibodies against human PHGPx showing the occurrence of PHGPx in platelets. (A) Polyclonal antibody against a peptide fragment of human PHGPx, (B) Monoclonal antibody against human PHGPx. Lysates of A431 cells and recombinant human PHGPx (mutant *sec46cys*) were used as positive and the lysate of erythrocytes as negative controls.

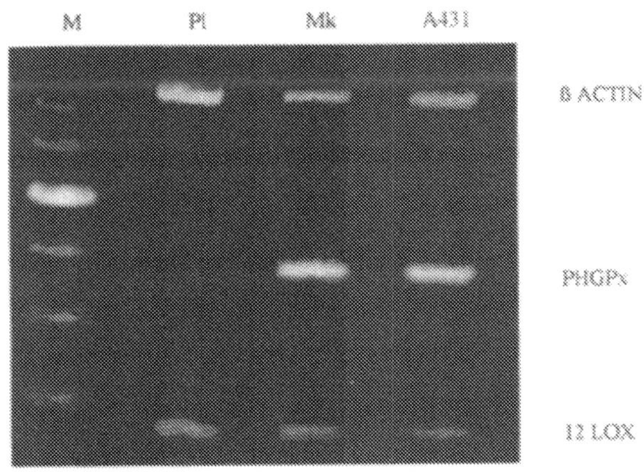

Figure 2. RT-PCR of mRNAs of PHGPx, 12-LOX and ß-actin in human platelets and megakaryocytes. Total RNA was reverse-transcribed using 5'-TGTGCGCGCTCCATGCACGAGT and 5'-AAATAGTGGGGCAGGTCCTTCTCT as primer set for PHGPx-cDNA and 5'-CTGGCCCCAGAAGATCTGATC and 5'-GATGATCTACCTCCAAATATG as primer set for platelet-type 12-LOX-cDNA. ß-Actin-cDNA served as biological standard. *Left to right:* molecular mass standard mixture (M), platelets (Pl), megakaryocytes (Mk). A431 cells (positive control).

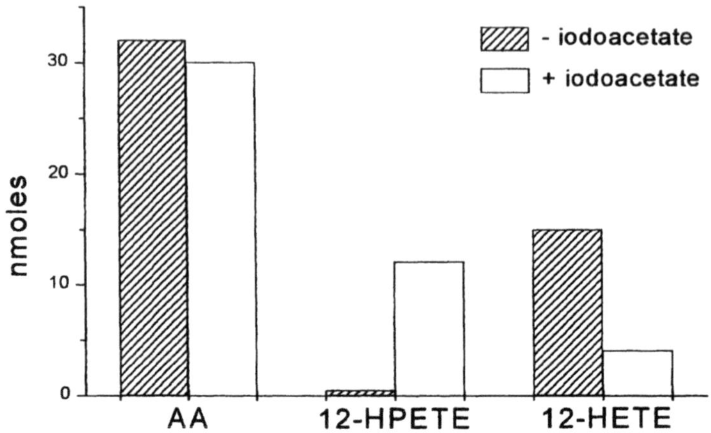

Figure 3. Amounts of AA metabolites in human platelets with and without pretreatment with iodoacetate. Washed human platelets with or without pretreatment with iodoacetate were incubated with 130 µM AA for 10 minutes. Following diethylether extraction the ethereal phase was dried down, reconstituted in HPLC-solvent and analyzed by HPLC on a C_{18}-column using n-hexane:2-propanol:acetic acid (100:2:0.1, v/v) as mobile phase and 13-HODE as an internal standard. For analysis of hepoxilins samples were derivatized with 9-anthranyl diazomethane [2]. Methanol:acetonitrile:water (90:6:4, v/v) was used as eluent in this case.

Acknowledgment

This study was generously supported by the Free University Berlin (Ni-FO97). Authors wish to thank Dr. A. Maurer, Department of Hematology, UKBF for providing bone marrow samples, Dr. H. Kühn, Institute of Biochemistry, HU Berlin for supplying PHGPx antibodies and Dr. C.R. Pace-Asciak, Research Institute, HSC Toronto for providing hepoxilins.

References

1. Yamamoto S, Suzuki H, Ueda N. Arachidonate 12-lipoxygenases. Prog Lipid Res 1997; 36:23-41.
2. Pace-Asciak CR, Reynaud D, Demin P, Nigam S. The hepoxilins, a review. In: Nigam S, Pace-Asciak CR, editors. Lipoxygenases and their metabolites. New York: Kluwer/Plenum, 1999: 123-132.
3. Brigelius-Flohé R. Tissue-specific functions of individual glutathione peroxidases. Free Radic Biol Med 1999;27:951-965.
4. Bryant RW, Simon TC, Bailey JM. Role of glutathione peroxidase and hexose

monophosphate shunt in the platelet lipoxygenase pathway. J Biol Chem 1982;257:14937-43.

5. Weitzel F, Wendel A. Selenoenzymes regulate the activity of leukocyte 5-lipoxygenase via the peroxide tone. J Biol Chem 1993;268:6288-92.
6. Schnurr K, Belkner J, Ursini F, Schewe T, Kühn H. The selenoenzyme phospholipid hydroperoxide glutathione peroxidase controls the activity of the 15-lipoxygenase with complex substrates and preserves the specificity of the oxygenation products. J Biol Chem 1996;271:4653-58.
7. Imai H, Narashima K, Arai M, Sakamoto H, Chiba N, Nakagawa Y. Suppression of leukotriene formation in RBL-2H3 cells that overexpressed phospholipid hydroperoxide glutathione peroxidase. J Biol Chem 1998;273:1990-97.
8. Schewe T, Rapoport SM, Kühn H. Enzymology and physiology of reticulocyte lipoxygenase: comparison with other lipoxygenases. Adv Enzymol Relat Areas Mol Biol 1986;58:191-272.
9. Maiorino M, Aumann KD, Brigelius-Flohé R, et al. Probing the presumed catalytic triad of selenium-containing peroxidases by mutational analysis of phospholipid hydroperoxide glutathione peroxidases (PHGPx). Biol Chem Hoppe Seyler 1995;376:651-60.
10. Weitzel F, Wendel A. Phospholipid hydroperoxide glutathione peroxidase in various mouse organs during selenium deficiency and repletion. J Biol Chem 1993;268:6288-92.
11. Sutherland M, Schewe T, Nigam S. Biological actions of the free acid of hepoxilin A_3 on human neutrophils. Biochem Pharmacol 1999;59:435-40.
12. Margalit A, Sofer Y, Grossman S, Reynaud D, Pace-Asciak CR, Livne AA. Hepoxilin A_3 is the endogenous lipid mediator opposing hypotonic swelling of intact human platelets. Proc Natl Acad Sci USA 1993;90:2589-92.

TRANSCRIPTIONAL REGULATION OF THE HUMAN LEUKOTRIENE C_4 SYNTHASE PROMOTER

Ji-liang Zhao, K. Frank Austen, and Bing K. Lam

The cysteinyl leukotrienes are a class of molecules originally known as the slow-reacting substances of anaphylaxis (SRS-A). Leukotriene (LT)C_4, the parent compound of the group, together with its derivatives LTD_4 and LTE_4 cause increased mucus secretion, vascular permeability, and smooth muscle constriction. The pathologic role of the cysteinyl leukotrienes in bronchial asthma has been established by their chronic overproduction in the airways of patients with asthma and by the relief experienced by such patients after the administration of agents that attenuate leukotriene elaboration or their receptor-mediated actions [1]. LTC_4 synthase (LTC_4S), an 18-kD integral membrane protein, is a key enzyme in the biosynthesis of the cysteinyl leukotrienes. It conjugates LTA_4 with reduced glutathione to form LTC_4. LTC_4S is expressed predominantly in mast cells, basophils, and eosinophils, a subset of the monocyte and myeloid lineages important for allergy and inflammation. However, the mechanisms controlling the cell-specific expression of LTC_4S are not well understood. Our recent study of the LTC_4S promoter indicated that several *cis*-acting elements are involved in the transcription of this TATA-less gene. This article reviews constitutive and induced expression of LTC_4S in various cells and focuses on the delineation of *cis*-acting elements and the identification of transcription factors that transactivate the human LTC_4S gene promoter.

Constitutive LTC_4S Expression Regulated by Lineage Differentiation

Among the proteins involved in the biosynthesis of the leukotrienes, cytosolic phospholipase A_2 ($cPLA_2$) is present in all cell types, while 5-lipoxygenase (5-LO) and 5-lipoxygenase activating protein (FLAP) are expressed prominently in cells of myelomonocytic lineage and to a much lesser extent in B cells. LTA_4 hydrolase is widely expressed in both nonhematopoietic and hematopoietic cells; among the latter, its expression relative to LTC_4S can be selective (human polymorphonuclear leukocytes), dominant (monocyte/macrophages), or minimal (mast cells). In contrast, the expression of LTC_4S is limited: LTC_4S is the only leukotriene pathway terminal enzyme for eosinophils and basophils, exhibits predominance in mast cells, and occurs in platelets in the absence of the proximal pathway enzymes so as to depend on a transcellular supply of LTA_4.

The restricted pattern of LTC_4S expression in certain related cell lineages indicates

B. Samuelsson et al. (eds.), Advances in Prostaglandin and Leukotriene Research, 31–36.

that lineage development and differentiation lead to the constitutive expression of LTC_4S. This developmental regulation has been documented during differentiation and maturation of human eosinophils from cord blood progenitors with interleukin (IL)-3 and IL-5 in the presence of a reconstituted basement membrane, Matrigel [2]. The cord blood mononuclear cells had no LTC_4S as measured by Western blot; cells of mixed granulocytes after 7 days in culture produced no LTC_4 when stimulated with ionophore, even though they expressed the proximal pathway proteins $cPLA_2$, 5-LO, and FLAP. As the cells became 93% pure hybrid granulocytes, containing both eosinophil and basophil granules after 14 days in culture, both LTC_4S transcript and protein were observed. The LTC_4S protein and activity continued to increase as the hybrid eosinophils lost basophilic granules to become mature eosinophils [2]. Thus, the data revealed a direct correlation between the developmental stages of eosinophils and the level of LTC_4S expression.

Mitogenic and Inflammatory Stimuli Induce LTC_4S Expression

LTC_4S expression can be induced in many leukemia cell lines, with levels ranging from none (HL-60 cells) to minimal (U937 cells) to high (HEL and THP-1 cells), without mitogenic stimulation. Because these cells are immortal and are arrested at particular developmental stages, they have been used as model systems to show the effect of terminal differentiation induced by various mitogenic stimuli on the expression of LTC_4S, a process that mimics the developmental regulation of LTC_4S in *in vitro*-derived eosinophils. For example, HL-60 cells, which underwent neutrophilic differentiation in presence of 1.2 % dimethyl sulfoxide (DMSO), expressed the transcript and proteins of LTA_4 hydrolase, but not those of LTC_4S [3]. However, when stimulated with 0.4 mM butyric acid, an HL-60 subline demonstrated eosinophilic differentiation, accompanied by a progressive induction of LTC_4S transcript and protein and an approximately 75-fold increase of LTC_4S activity in 7 days [3].

Similarly, LTC_4S activity increased in U937 cells stimulated for 3 days by 1.3% DMSO, but not by 10 nM phorbol myristate acetate (PMA) [4]. However, PMA at a 320 nM concentration induced LTC_4S activity in HEL cells [5]. Thus, LTC_4S can be upregulated in leukemic cell lines by appropriate terminal differentiation through exogenous biochemical stimulation. Cytokines play a major role in stimulating LTC_4S expression in inflammatory settings. A proinflammatory cytokine, such as transforming growth factor β-(TGF_β), rapidly increases LTC_4S transcript in THP-1 cells [6]. Besides IL-3 and IL-5, which are important regulators of eosinophil development and survival, granulocyte-macrophage colony-stimulating factor (GM-CSF) can enhance the transition of normodense eosinophils to a hypodense form, which generates increased amounts of LTC_4 when stimulated with FMLP [7]. In bronchial biopsy specimens from patients with aspirin-intolerant asthma (AIA), both the number of eosinophils in the airways and the number of cells with LTC_4S protein expression were increased as measured by immunohistochemistry [8]. The upregulation of LTC_4S protein expression in AIA is likely to be a maturation effect on eosinophils by a network of cytokines released from inflammatory cells, including T cells.

Characterization of the LTC_4S Promoter

Although an increase in steady-state LTC_4S transcripts has been observed during eosinophil development or appropriate terminal differentiation of leukemic cells, it is not clear whether the increase is the result of LTC_4S gene transcription or transcript stability. So far, no nuclear run-on experiment to directly demonstrate that LTC_4S is regulated at transcription level has been reported. The half-life of LTC_4S transcript from TGF_β-treated THP-1 cells did not significantly differ from that of untreated cells, suggesting that either an unknown transcript stability mechanism or an increase in transcription rate was involved in upregulation of LTC_4S mRNA in TGF_β-treated THP-1 cells [6]. When a 1.7-kb LTC_4S promoter/luciferase reporter construct was transiently transfected into THP-1 cells, the promoter activity was increased by less than 2-fold in response to TGF_β. In comparison, TGF_β increased the steady-state LTC_4S mRNA in THP-1 cells by 7-fold in 24 hours [6]. The data suggest that LTC_4S mRNA levels may be regulated by a complex mechanism involving both transcription and transcript stability. In addition, LTC_4S can be regulated post-translationally by phosphorylation, which suppresses LTC_4S activity through a protein kinase C–dependent pathway [9].

Like the 5-LO and LTA_4 hydrolase gene promoters, the LTC_4S promoter contains no TATA or CAAT box. The transcription start site of the LTC_4S gene has previously been mapped by primer extension to nucleotides –66, –69, and –96 with respect to the translation start site (+1) with poly $(A)^+$ RNA from *in vitro*-derived eosinophils and KG-1 cells [10]. A single transcription start site at –78 was also reported with total RNA from THP-1 cells [11]. Sequence analysis of the 5' region reveals the presence of multiple consensus sequences, including binding sites for AP2 (–973), AP1 (–903), and Sp1 (–118). More recently, a polymorphism of an A to C transversion at –444 was identified in the 5' flanking region of the LTC_4S gene. However, the single nucleotide polymorphism appears to associate with the AIA phenotype only in eastern Europe and not in the United States [12,13]. Furthermore, the promoter activity of a construct containing the mutation is not increased when the construct is transfected into THP-1 cells as compared to the promoter activity associated with a wild-type construct [13].

We cloned 2.1 kb of the human LTC_4S 5' flanking region, which contains the three DNase I hypersensitivity sites, ahead of a promoterless/enhancerless luciferase reporter gene and transiently transfected the construct and its variants into THP-1 and K562 cells. As THP-1 cells express LTC_4S and the K562 cells do not, the results were informative for cell-specific expression. Deletional analysis indicated that two regions, from –557 to –228 and from –175 to –122, respectively, were important for the promoter activities of the construct in THP-1 cells but not in K562 cells [14]. Thus, these regions are likely to be involved in the cell-specific expression of the promoter. Further analysis of the proximal cell-specific region by linker-scanning mutagenesis revealed that a 40-bp sequence from –164 to –125 contains positive *cis*-elements for cell-specific reporter activity in transfected

THP-1 cells. These elements include an AP3 site, a tandem CACCC repeat, and a CCCTC motif; mutation of any of these sites caused the promoter activity to decrease by 35 to 75% of that of the wild-type construct. Deletional analysis also revealed a non-cell-specific basal promoter region active in both THP-1 and K562 cells. This region contains a GC-box (–120 to –115) and an initiator (Inr) motif (CAGAC at –66 to –62), which when mutated singularly or together, reduced the reporter gene expression by 60% and 80%, respectively [14].

Transactivation of the LTC_4S Promoter by Sp1 and Kruppel-like Factors

DNA-protein complexes were observed in electrophoretic mobility shift assays (EMSA) with ^{32}P-labeled oligonucleotides, containing the individual motif of the *cis*-elements identified as basal and cell-specific, respectively. EMSA of the AP3 site and of the Inr sequence each yielded a single gel shift band with nuclear extract isolated from THP-1 and K562 cells, and the complexes were blocked by their respective excess cold oligonucleotides [14]. However, the proteins binding to the AP3 site were not sought because the AP3 binding candidates are not yet reported. A novel Inr binding protein appeared to recognize the LTC_4S Inr motif (CAGAC), which is different from other Inr motifs, such as $CA_{+1}TTC$ (the classic terminal deoxynucleotidyl transferase gene Inr), CATCTT, CACTC, and CATTT, the binding sites for TFIID, TFII-I, USF, and YY1, respectively [15-18].

Multiple EMSA complexes were found with THP-1 cell nuclear extract and a ^{32}P-labeled oligonucleotide containing the CCCTC site, which overlaps the proximal CCCAC repeat sequence. Interestingly, one of the complexes observed with THP-1 nuclear extract was absent when K562 nuclear extract was used in EMSA, indicating that the binding protein was expressed specifically in THP-1 cells and could be involved in the cell-specific expression of LTC_4S. The transcription factor Sp1 family members, Sp1 and Sp3, were found to bind to the GC-box (–120 to –115) in the LTC_4S basal promoter region as well as the distal CACCC motif (–149 to –145), as determined by super-shift analysis with antisera against individual Sp1 family members [14]. Besides Sp1 and Sp3, the distal CACCC motif was also found in EMSA to bind recombinant erythroid Kruppel-like factor (EKLF) protein, which is the founding member of the Kruppel-like family of transcription factors [19]. The EKLF family proteins recognize a core sequence of CACCC and transactivate many cell-specifically expressed and/or inducible genes. Therefore, Zf9/CPBP (core promoter-binding protein) [20], a subfamily member of the EKLF family was used in cotransfection experiments to determine whether a Kruppel-like factor is involved in LTC_4S gene transcription. When cotransfected into COS cells, Zf9/CPBP transactivated the reporter activity of the wild-type –228 to –3 LTC_4S promoter construct by 4-5 fold, but not its variant with mutations in the CACCC repeat region. Similarly, Sp1 transactivated the LTC_4S –228 to –3 promoter in *Drosophila Schneider* cells, which lack endogenous Sp1 family transcription factors. The transactivation of LTC_4S promoter activity by Sp1 was reduced by about 67% with mutations in the distal CACCC motif [14].

The proposed binding of *cis*-elements by their putative transcription factors so as to direct LTC_4S transcription in THP-1 cells is schematically depicted in Figure 1. The CACCC repeats play an important role in regulating the cell-specific expression of LTC_4S by interacting with at least two families of transcription factors, Sp1 and Sp3 and the Kruppel-like factors. Sp1 is involved in transcription of many TATA-less genes, including the 5-LO gene, which contains five tandem repeats of a GC-rich sequence in its proximal promoter region. The region binds to Sp1 and Egr-1, both of which transactivated the 5-LO promoter in *Drosophila Schneider* cells in separate cotransfections [21,22], whereas only Sp1 and Sp3 were active with the LTC_4S promoter [14]. In the LTC_4S promoter, the transcription factors that bind to the CACCC repeats may interact with factors binding to other regions, including the CCCTC site and the GC-box, to achieve the cell-specific effect of LTC_4S gene regulation.

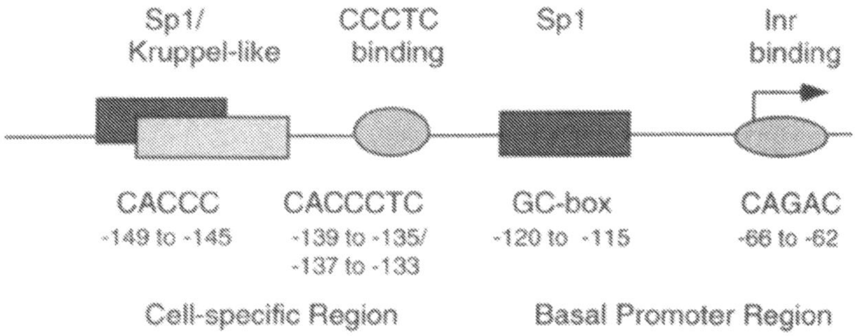

Figure 1. Schematic presentation of the transcriptional regulation of the LTC_4S proximal promoter. The sequences of the LTC_4S *cis*-elements are listed and numbered with respect to the translation start site (+1). The arrow indicates the transcription start site at −66 as previously identified. The known transcription factors that bind to the *cis*-elements in the LTC_4S proximal promoter are depicted as boxes (dark- and light-shaded for Sp1 and Kruppel-like factor, respectively), and the putative binding proteins are depicted as ovals.

References

1. Drazen JM, Israel E, O'Byrne PM. Treatment of asthma with drugs modifying the leukotriene pathway. N Engl J Med 1999;340:197-206.
2. Boyce JA, Lam BK, Penrose JF, et al. Expression of LTC_4 synthase during the development of eosinophils in vitro from cord blood progenitors. Blood 1996;88:4338-47.
3. Scoggan KA, Nicholson DW, Ford-Hutchinson AW. Regulation of leukotriene-biosynthetic enzymes during differentiation of myelocytic HL-60 cells to eosinophilic or neutrophilic cells. Eur J Biochem 1996;239:572-78.
4. Nicholson DW, Ali A, Klemba MW, Munday NA, Zamboni RJ, Ford-Hutchinson AW. Human leukotriene C_4 synthase expression in dimethyl sulfoxide-differentiated U937 cells. J Biol Chem 1992;267:17849-57.

5. Soderstrom M, Bolling A, Hammarstrom S. Induction of leukotriene C_4 synthase activity in differentiating human erythroleukemia cells. Biochem Biophys Res Commun 1992;189:1043-49.
6. Riddick CA, Serio KJ, Hodulik CR, Ring WL, Regan MS, Bigby TD. TGF-beta increases leukotriene C_4 synthase expression in the monocyte-like cell line, THP-1. J Immunol 1999;162:1101-7.
7. Owen WF, Jr., Petersen J, Austen KF. Eosinophils altered phenotypically and primed by culture with granulocyte/macrophage colony-stimulating factor and 3T3 fibroblasts generate leukotriene C_4 in response to FMLP. J Clin Invest 1991;87:1958-63.
8. Cowburn AS, Sladek K, Soja J, et al. Overexpression of leukotriene C_4 synthase in bronchial biopsies from patients with aspirin-intolerant asthma. J Clin Invest 1998;101:834-46.
9. Ali A, Ford-Hutchinson AW, Nicholson DW. Activation of protein kinase C down-regulates leukotriene C_4 synthase activity and attenuates cysteinyl leukotriene production in an eosinophilic substrain of HL-60 cells. J Immunol 1994;153:776-88.
10. Penrose JF, Spector J, Baldasaro M, et al. Molecular cloning of the gene for human leukotriene C_4 synthase. Organization, nucleotide sequence, and chromosomal localization to 5q35. J Biol Chem 1996;271:11356-61.
11. Bigby TD, Hodulik CR, Arden KC, Fu L. Molecular cloning of the human leukotriene C_4 synthase gene and assignment to chromosome 5q35. Mol Med 1996;2:637-46.
12. Sanak M, Simon HU, Szczeklik A. Leukotriene C_4 synthase promoter polymorphism and risk of aspirin-induced asthma. Lancet 1997;350:1599-600.
13. Van Sambeek R, Stevenson DD, Baldasaro M, et al. The 5' flanking region polymorphism of the gene encoding leukotriene C_4 synthase does not correlate with the aspirin-intolerant asthma phenotype in the United States. J Allergy Clin Immunol 2000; in press.
14. Zhao JL, Austen KF, Lam BK. Cell-specific transcription of leukotriene C_4 synthase involves a Kruppel-like transcription factor and Sp1. J Biol Chem 2000;275:8903-10.
15. Javahery R, Khachi A, Lo K, Zenzie-Gregory B, Smale ST. DNA sequence requirements for transcriptional initiator activity in mammalian cells. Mol Cell Biol 1994;14:116-27.
16. Manzano-Winkler B, Novina CD, Roy AL. TFII is required for transcription of the naturally TATA-less but initiator-containing Vbeta promoter. J Biol Chem 1996;271:12076-81.
17. Du H, Roy AL, Roeder RG. Human transcription factor USF stimulates transcription through the initiator elements of the HIV-1 and the Ad-ML promoters. EMBO J 1993;12:501-11.
18. Seto E, Shi Y, Shenk T. YY1 is an initiator sequence-binding protein that directs and activates transcription in vitro. Nature 1991;354:241-45.
19. Miller IJ, Bieker JJ. A novel, erythroid cell-specific murine transcription factor that binds to the CACCC element and is related to the Kruppel family of nuclear proteins. Mol Cell Biol 1993;13:2776-86.
20. Ratziu V, Lalazar A, Wong L, et al. Zf9, a Kruppel-like transcription factor up-regulated in vivo during early hepatic fibrosis. Proc Nat Acad Sci U S A 1998;95:9500-5.
21. In KH, Asano K, Beier D, et al. Naturally occurring mutations in the human 5-lipoxygenase gene promoter that modify transcription factor binding and reporter gene transcription. J Clin Invest 1997;99:1130-37.
22. Silverman ES, Du J, De Sanctis GT, et al. Egr-1 and Sp1 interact functionally with the 5-lipoxygenase promoter and its naturally occurring mutants. Am J Respir Cell Mol Biol 1998;19:316-23.

Localization and Translocation of Enzymes Involved in Leukotriene Biosynthesis

Yoshiyuki Ide, Hiromi Hanaka, Tetsuya Hirabayashi, Takao Shimizu, and Takashi Izumi

Introduction

Leukotrienes (LT) are potent proinflammatory mediators which are thought be released immediately upon extracellular stimulation. Responding to stimulation, arachidonic acid (AA) is released from phospholipids by the action of phospholipases ($cPLA_2$, $sPLA_2$, etc.). Once liberated form the phospholipid, AA is further metabolized to LTA_4 by 5-lipoxygenase (5-LO). LTA_4 hydrolase (LTA_4H) catalyzes the transformation of LTA_4 to LTB_4, and LTC_4 synthase adds glutathione to LTA_4 and produces LTC_4, which is subsequently transformed to LTD_4 and LTE_4 by gamma-GTP and dipeptidase, respectively and sequentially [1].

Figure 1 shows the localization of enzymes involved in LT production before and after stimulation. Though the intracellular localization of each enzyme in resting cells may be varied in each type of cells [2,3], the figure shows the typical localization. In resting cells, both FLAP and LTC_4 synthase are localized around the nuclear membrane and partially in ER, while others are mainly localized in the cytosol. 5-LO is observed both in the nucleus and the cytosol depending on cell types. It has been reported that once cells are activated to produce LT, $cPLA_2$ and 5-LO immediately accumulate around the nucleus [2]. But the mechanism and the biological significance are not well known. Here we report the translocation of these enzymes.

Intracellular Translocation of $cPLA_2$ and AA Release

Many studies indicated that $cPLA_2$ play important roles in the production of lipid mediators. We observed translocation of $cPLA_2$ after stimulation with PAF in CHO cells expressing both the PAF receptor and a fusion protein of $cPLA_2$ and EGFP [4] (Figure 2). In the resting cells, the fluorescent signal was observed equally in the cytosol, but the strong fluorescent signal appeared around the nucleus from one to five minutes after PAF stimulation. This translocation was modified differently by adding EGTA to the chelate medium calcium ion at the point of one or five minutes after stimulation with PAF. When EGTA was added at one minute after PAF stimulation, the $cPLA_2$ gathered around the

B. Samuelsson et al. (eds.), Advances in Prostaglandin and Leukotriene Research, 37–40.

nucleus translocated to the cytosol immediately, whereas in the cells to which EGTA was added at the five-minute point, the $cPLA_2$ remained localized around the nucleus. Then, to clarify the correlation between this calcium-responding translocation and $cPLA_2$ activity, we measured release of [^{3}H] AA from the CHO cells with PAF stimulation. After PAF stimulation, 8% of total cell [^{3}H] AA was extracellularly released. The release was inhibited to 2%, the same level as the control with calcium chelation at one minute. On the other hand, when EGTA was applied five minutes after stimulation, 8% of AA was released. These data suggest that the translocation and succeeding localization around the nucleus over 5 minutes is important to derive AA from the membrane.

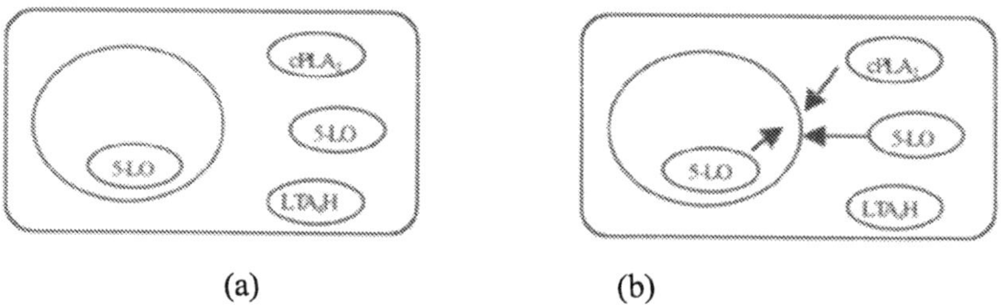

Figure 1. Localization of enzymes in a) resting cells and b) activated cells.

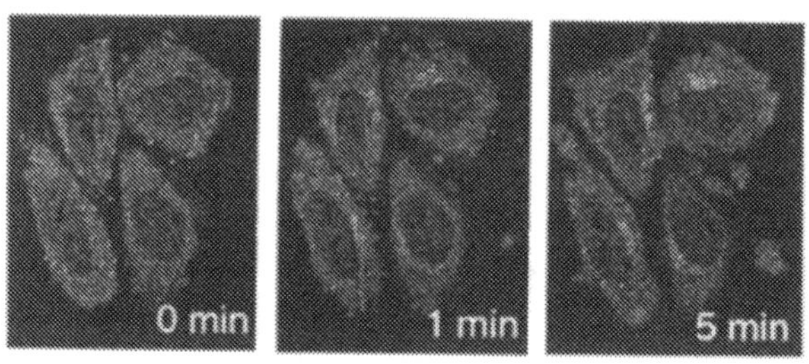

Figure 2. Localization of cPLA and EGFP fusion protein at each time after the stimulation with PAF.

Translocation and Localization of 5-LO and LTA_4H

Then we prepared CHO cells stably expressing 5-LO and EGFP fusion protein and observed the movement of 5-LO triggered by calcium ionophore (Figure 3). In resting cells 5-LO was localized in both the nucleus and the cytosol (Figure 3a). This nuclear localization was found to be regulated by nuclear localization signal using 5-LO differently mutated on a putative NLS. Moreover, under the treatment of calcium ionophore for 10 minutes, the enzyme translocated around the nucleus sharply (Figure 3b).

We also observed the localization of LTA_4H in rat alveolar macrophages (AMs) with immunocytochemistry (Figure 4) and CHO cells stably expressing LTA_4H and EGFP fusion protein (data not shown). Freshly collected AMs from BALF showed LTA_4H was localized mainly to the cytosol and after floating for 10 minutes the enzyme partially translocated into the nucleus. This localization and translocation were also observed in CHO cells. The translocation was inhibited when CHO cells were treated with tyrosine kinase inhibitors, or when FCS was starved from the medium. The localization of LTA_4H was also changed 5 minutes after adhesion on a slide glass and the fluorescent signal was observed around the plasma membrane. Different from $cPLA_2$ and 5-LO, LTA_4H did not translocate when cells were treated with calcium ionophore. These data indicate LTA_4H could also change the localization from cytosol to nucleus or plasma membrane coordinating with adhesion but did not depend on the intracellular calcium increase.

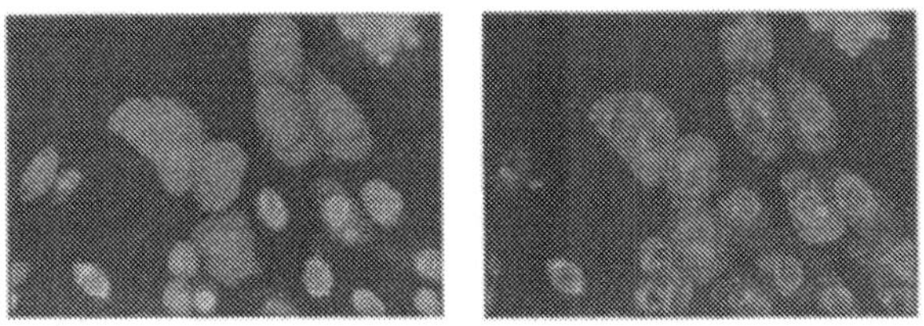

(a) (b)

Figure 3. Localization of 5-LO in CHO cells. Localization of 5-LO and EGFP fustion protein in CHO cells a) before and b) 10 minutes after stimulation with calcium ionophore.

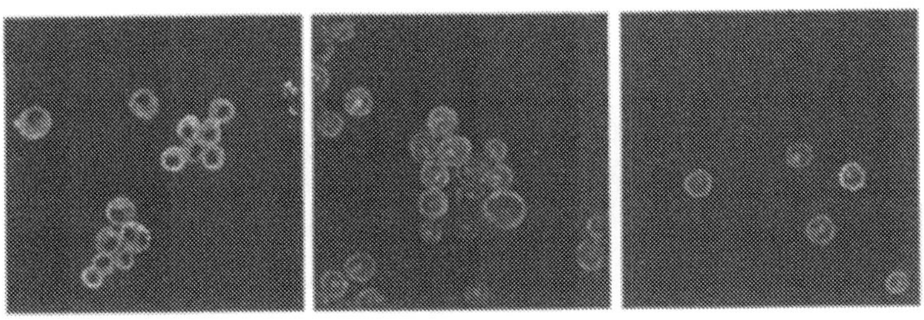

(a) (b) (c)

Figure 4. Localization of LTA_4H in AMs. LTA_4H was visualized using antibody in AMs, a) in BALF, b) kept floating for 30 minutes, c) adhered on a slide glass for 5 minutes.

Summary

Here we sequentially observed translocation of $cPLA_2$, 5-LO, and LTA_4H in CHO cells expressing each enzyme stably, or in AMs. 5-LO and $cPLA_2$ translocated to the nuclear membrane upon stimulation increasing intracellular calcium, and it also appeared that transient calcium increase was not enough for $cPLA_2$ to produce AA. LTA_4H also translocated to the plasma membrane not when intracellular calcium increased but AMs adhered on a slide glass. The meaning of this translocation is not clear yet, but these data may suggest that localization of these enzymes might regulate the production of LT.

Acknowledgement

This work was supported in part by a grant-in-aid from the Ministry of Education, Science and Culture of Japan.

References

1. William S. The eicosanoids and their biochemical mechanisms of action. Biochem J 1989;259;315-24.
2. Peters-Golden M, McNish R.W Redistribution of 5-lipoxygenase and cytosolic phospholipase A_2. Biochem Biophys Res Commun 1993;196;147-53.
3. Channon J, Leslie C. A calcium-dependent mechanism for associating a soluble arachidonoyl- hydrolyzing phospholipase A_2 with membrane in the macrophage cell line RAW 264.7. J Biol Chem 1990;265;5409-13.
4. Hirabayashi T, Kume K, Hirose K. Critical duration of intracellular Ca^{2+} response required for continuous translocation and activation of cytosolic phospholipase A_2 J Biol Chem 1999;274;5163-69.

Regulation of COX-2 Expression in Fibroblasts, Osteoblasts, Mast Cells, and Macrophages

Harvey R. Herschman, David J. Wadleigh, and Srinivasa T. Reddy

Introduction

The prostaglandin synthase/cyclooxygenase (COX) enzymes catalyze two reactions. The cyclooxygenase activity of these enzymes first converts free arachidonic acid, released from membrane lipids by a phospholipase, to PGG_2. This intermediate is then converted to PGH_2 by the COX hydroperoxidase activity. PGH_2 is the precursor for all prostanoids–prostaglandins, thromboxanes and prostacyclins.

Cyclooxygenase activity is present in nearly all cells. For many years the rate-limiting step in prostanoid production in response to cell stimulation was thought to be activation of phospholipase(s) to release membrane-bound arachidonic acid. Prostaglandin production was thought to be limited by arachidonic acid availability; constitutive levels of COX were thought to be present in excess in cells, and available to convert free arachidonic acid to PGH_2. It was assumed that only one COX gene is present in mammalian genomes. However, studies of glucocorticoid inhibition of prostaglandin production [1], antigenic isoforms of COX [2], and identification of an inducible mRNA that cross-hybridized with "COX" cDNA [3] suggested a second, inducible COX gene might exist.

We identified "immediate-early" genes induced by the mitogen/tumor promoter tetradecanoyl phorbol acetate (TPA), and named these cDNAs TPA-Induced-Sequences, or TIS genes [4]. TIS10 encodes a protein with sequence similarity to the previously cloned murine cyclooxygenase [5]. Recombinant TIS10 protein has both cyclooxygenase and hydroperoxidase activities; TIS10 encodes a second COX [6]. The "constitutive" COX is now referred to as COX-1; the inducible (TIS10) form is termed COX-2. Xie et al. [7] cloned a cDNA from chicken fibroblasts that encodes a protein with strong sequence homology to mammalian COX-1. However, they could not determine whether CEF-147 was a second cycloxygenase or the avian COX-1 homologue, since avian COX-1 had not been cloned. O'Banion et al. later cloned COX-2 as a serum-induced gene [8,9].

Regulation of COX-2 Gene Expression in Fibroblasts

We characterized the regulatory regions of the murine COX-2 gene, the transcription factors, and the signal transduction pathways that mediate COX-2 induction by the v-src oncogene and

B. Samuelsson et al. (eds.), Advances in Prostaglandin and Leukotriene Research, 41–47.

by platelet-derived growth factor (PDGF) in fibroblasts. Using deletion analysis, with chimeric luciferase reporter constructs, we identified the region of the COX-2 promoter required for v-src [10] and PDGF induction [11]. Gel shift analyses identified proteins that bind to the overlapping cyclic AMP response element (CRE) and Ebox elements present at nucleotides -56 to -48 of the COX-2 gene [12]. Mutational analysis identified the CRE as the critical cis-acting element regulating both v-src [10] and PDGF [11] induction. Using cotransfection assays with plasmids encoding dominant-negative and/or constitutively activated forms of Ras, MAP kinase and MEKK, we demonstrated (i) that the key transcription factor mediating both v-src and PDGF induction via the CRE of the COX-2 promoter is c-Jun, and (ii) that both v-src and PDGF induction are mediated by the Ras/MEKK1/JNK/Jun signal transduction pathway [10,11].

Regulation of COX-2 Gene Expression in Osteoblasts

Osteoblasts play a major role in the prostaglandin synthesis that modulates bone resorption. The MC3T3-E1 line, established from the calvaria of a newborn mouse, can differentiate into osteoblast-like cells. When stimulated by a wide variety of ligands that include bFGF, PDGF, PGE_2 or a combination of (TNFα and IL1β), differentiated MC3T3-E1 cells produce prostaglandin E_2. We investigated COX-2 induction in MC3T3-E1 cells by these ligands, to determine the nature of the signaling pathways, transcription factors and cis-acting elements of the COX-2 gene that modulate COX-2 expression by alternative ligands in a common cell type.

We initially examined the cis-acting regulatory elements in COX-2 induction by bFGF in MC3T3-E1 osteoblasts [13]. We constructed wild-type and mutant luciferase reporter constructs containing 724 bp of the COX-2 gene 5' flanking sequence, and created mutations in a presumptive NFκB site, two potential NF-IL6 (C/EBP) sites, the CRE site and the Ebox. We examined luciferase levels following transient transfection of MC3T3-E1 cells and induction by bFGF. Mutation of the Ebox or the NFκB site has no effect. Mutation of either NF-IL6 site alone reduces reporter activity only slightly. However, mutation of both NF-IL6 sites reduces luciferase induction by bFGF nearly 70%. Mutation of the COX-2 CRE exhibits the strongest effect; no induction is observed. The CRE site, essential for both oncogene and PDGF induction in fibroblasts, is also essential for induction by serum, bFGF, PDGF, PGE_2 and (TNFα + IL1β) in osteoblasts. Deleting both NF-IL6 sites also impairs induction by all these ligands [13].

To investigate the role of the MEKK/JNK signaling pathway in osteoblasts, we cotransfected MC3T3-E1 cells with the COX-2 reporter, along with vectors expressing dominant negative (DN) forms of JNK and MEKK. Induction of luciferase activity by bFGF, PDGF, PGE_2, and (TNFα + IL1β) is substantially repressed. We conclude that the MEKK/JNK signaling pathway mediates induction of the COX-2 gene in osteoblasts, just as it does for oncogene and PDGF induction in fibroblasts [13].

Transcription from a CRE is usually activated by CREB or ATF transcription factors. However, COX-2 expression in fibroblasts, in response to v-src, serum or PDGF, is mediated by c-Jun activation at the CRE [10,11]. Overexpression of a c-Jun expression plasmid along with the wild-type COX-2 reporter enhances COX-2 reporter induction by serum, bFGF,

PDGF, PGE_2, or (TNFα + IL1β) in osteoblasts. In contrast, CREB overexpression does not enhance COX-2 reporter activity. Instead, wild-type CREB acts like a dominant negative; COX-2 reporter activation by serum bFGF, PDGF, PGE_2, or (TNFα + IL1β) is repressed by CREB overexpression. We conclude that c-Jun, not CREB, mediates induced COX-2 reporter expression in osteoblasts [13].

Transcription from NF-IL6 sites is generally mediated by C/EBP transcription factors. Since mutation of the COX-2 NF-IL6 sites reduces COX-2 promoter activation by all inducers in MC3T3-E1 osteoblasts, we cotransfected a COX-2 reporter along with expression vectors that encode wild-type and DN-C/EBP. Over-expression of C/EBP enhances COX-2 induction in MC3T3-E1 cells, while DN-C/EBP suppresses COX-2 reporter activity by serum, bFGF, PDGF, PGE_2, or (TNFα + IL1β), suggesting that C/EBP transcription factors mediate COX-2 induction in osteoblasts [13].

We conclude that the CRE and NF-IL6 sites of the COX-2 promoter are important for transcriptional induction by a number of stimuli in MC3T3-E1 osteoblasts. Transcriptional activation via the CRE, for all these agents, involves the MEKK/JNK signaling pathway, activating the c-Jun transcription factor at the CRE. C/EBP family proteins modulate transcriptional activation at the NF-IL6 sites.

Regulation of COX-2 Gene Expression in Mast Cells

Mast cells, activated by aggregation of IgE receptors, release PGD_2. Prostaglandin production in activated mast cells occurs in two phases; immediate PGD_2 release, completed within 10 minutes, and delayed PGD_2 synthesis/secretion. The immediate phase is due to conversion of arachidonic acid to prostaglandin by preexisting COX-1. The delayed phase requires activation-induced COX-2 production [14].

To investigate the cis-acting regions of the COX-2 gene that regulate induction in mast cells, MMC-34 murine mast cells were transfected with the same COX-2 reporter mutants used in the osteoblast studies [15]. Mutating the COX-2 CRE reduces activation-induced luciferase expression by more than 90%. While mutations in either the Ebox or NFκB elements do not inhibit induction, mutation of both NF-IL6 sites blocks COX-2 activation. As in fibroblasts [10,11] and osteoblasts [13], the CRE and NF-IL6 sites play critical roles in induced COX-2 expression in activated mast cells.

Over-expression of DN-Ras, DN-MEKK1, and DN-JNK block luciferase induction from the wild-type COX-2 promoter in activated MMC-34 mast cells, suggesting that Ras/MEKK1/JNK signaling mediates activation-induced COX-2 gene expression. Like the MEKK1/JNK pathway, the Raf/MEK/ERK pathway is required for COX-2 induction in fibroblasts [10,11]. Following aggregation of IgE receptors in mast cells, MEK and the ERKs are Raf-dependent targets of Ras activation. DN-Raf-1, DN-ERK1, and DN-ERK2 block COX-2 induction in activated MMC-34 cells; the Ras/Raf/MEK/ERK pathway is also necessary for COX-2 induction in mast cells [15].

In both fibroblasts [10,11] and osteoblasts [13], COX-2 induction from the CRE is mediated by c-Jun and not by CREB. We used overexpression of these transcription factors to determine if c-Jun is also the key player at the CRE in activated mast cells. Overexpressed CREB blocks luciferase expression induced from the COX-2 promoter by IgE receptor aggregation in MMC-34 cells. In contrast, c-Jun overexpression augments induced COX-2 expression in activated MMC-34 cells by more than 7-fold.

The data suggest that, as in fibroblasts and osteoblasts, c-Jun mediated induction via the CRE plays a major role in COX-2 expression in activated mast cells. As in osteoblasts [13], mutation of both NF-IL6 sites reduces inducibility of the COX-2 promoter in mast cells [15], suggesting a role for C/EBP transcription factors. C/EBP over-expression enhances luciferase expression in activated mast cells, while DN-C/EBP over-expression blocks induction from the COX-2 promoter in mast cells [15].

We conclude that activation of the COX-2 gene in mast cells is very similar to activation of the gene in fibroblasts; following receptor aggregation, Ras-dependent pathways activate c-Jun dependent transcription at the COX-2 CRE. Additional Ras-dependent events involving MEKK1 and ERK are required for induction, as are C/EBP-dependent transcriptional events occurring at the NF-IL6 sites of the COX-2 promoter. The activation of the COX-2 gene by aggregation of mast cell IgE receptors is shown in schematic form in Figure 1.

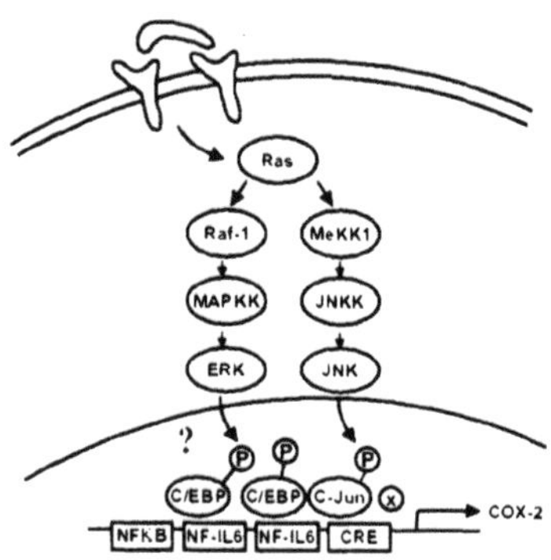

Figure. 1. The pathway of COX-2 induction in mast cells in response to aggregation of IgE receptors.

Regulation of COX-2 Gene Expression in Macrophages

Macrophages secrete prostaglandins in response to bacterial endotoxin/lipo-polysaccharide (LPS). Prostaglandin production in response to LPS is due primarily to induced COX-2 gene expression. COX-2 mediated prostaglandin production in macrophages plays a major role in acute and chronic inflammatory responses.

The same COX-2/luciferase constructs used to study COX-2 induction in osteoblasts [13] and mast cells [15] were transiently transfected into RAW 264.7 murine macrophages and the cells were induced with LPS. Mutation of the Ebox does not effect LPS induction. Mutation of both NF-IL6 sites strongly represses LPS induction from the COX-2 promoter in macrophages [16], as it does in osteoblasts and mast cells. As in fibroblasts [10,11], osteoblasts [13] and mast cells [15], mutation of the COX-2 CRE severely represses COX-2 induction in RAW 264.7 macrophages [16].

We used over-expression of wild-type and dominant-negative transcription factors to identify the mediators of LPS induction at the COX-2 NF-IL6 and CRE sites in macrophages. DN-C/EBP blocks LPS induced expression from the COX-2 promoter in RAW 264.7 cells, while over-expression of C/EBP enhances this induction; C/EBP transcription factors play a major role in LPS COX-2 induction in macrophages. CREB over-expression blocks luciferase induction from the COX-2 promoter in LPS-treated RAW 264.7 cells, while c-Jun over-expression enhances COX-2 reporter activation. In fibroblasts, osteoblasts, mast cells, and macrophages, c-Jun activation plays a role in the activation of the COX-2 promoter. In all cases, CREB does not seem to play a role.

Others have suggested that NFκB modulates induction of COX-2 by endotoxin in macrophages. However, mutation of the NFκB site has no effect on LPS induction of the COX-2 promoter/luciferase construct . Moreover, co-transfection of nondegradable I-κBα, which blocks LPS induction of an NFκB-luciferase reporter, does not block LPS-induced activation of the COX-2 promoter [16]. We do not, therefore, find evidence for NFκB mediated induction of the COX-2 gene in activated macrophages.

We anticipated that activation of c-Jun in LPS-treated macrophages would occur – as it does in fibroblasts, osteoblasts and macrophages – via the Ras/MEKK1/JNK pathway. As expected, DN-JNK or DN-MEKK1 significantly repress LPS activation of the COX-2 reporter in RAW 264.7 macrophages. However, expression of DN-Ras had no effect on LPS-induced luciferase activity in RAW 264.7 cells; activation of the JNK/MEKK signaling pathway and LPS-induced COX-2 transcription does not require Ras activity in LPS-treated RAW 264.7 macrophages, in contrast to the signaling pathways that activate c-Jun in fibroblasts, osteoblasts, and mast cells.

Unlike signaling requirements in fibroblasts and mast cells, Raf/MAPKK/ERK signaling also does not play a role in LPS-induced COX-2 gene expression in activated macrophages. Dominant negative forms of these molecules do not block luciferase induction from the COX-2 promoter in LPS-treated macrophages [16].

How does the LPS-induced signal get to c-Jun? TLR4 is the major LPS receptor. ECSIT (evolutionarily conserved signaling intermediate in toll pathways) was identified as an adapter protein that bridges toll/tumor necrosis factor receptor associated factor (TRAF6) activation to MEKK1 and facilitates MEKK1 activation of c-Jun [17]. We thought ECSIT might couple LPS activation to MEKK/JNK/c-Jun-dependent COX-2 gene expression in macrophages. Cotransfection with a plasmid expressing a DN-ECSIT protein significantly represses LPS-induction of COX-2 reporter activity in macrophages, suggesting that LPS-activated macrophages signal to the MEKK1/JNK pathway from ligand-bound LPS receptors through ECSIT, rather than Ras [17]. The activation of the COX-2 gene by endotoxin treatment of macrophages is shown on schematic form in Figure 2.

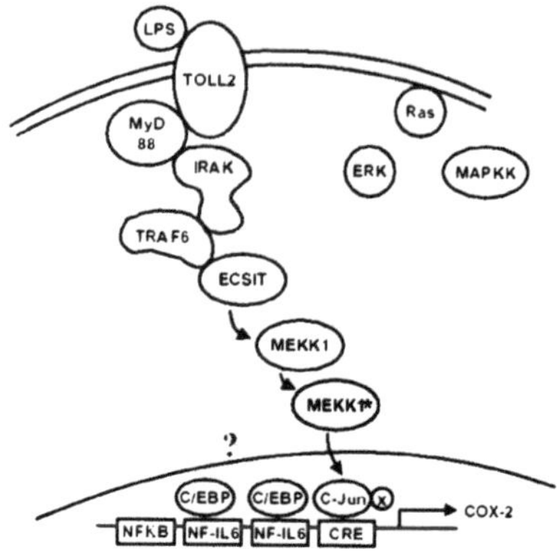

Figure 2. The pathway of COX-2 induction in macrophages in response to endotoxin stimulation.

Conclusions

The COX-2 promoter CRE and NF-IL6 sites play major roles in COX-2 induction in murine cells, in response to oncogenes, hormones, cytokines, and growth factors. In all cases examined, c-Jun appears to be the transcription factor modulating transcription from the CRE. C/EBP proteins are active at the NF-IL6 sites. c-Jun activation in mast cells and macrophages occurs by distinct signaling pathways, suggesting alternative targets for pharmacologic intervention of COX-2 expression in distinct cell types.

Acknowledgments

Supported by NIH Asthma, Allergic and Immunologic Disease Center Grant AI34567.

References

1. Masferrer JL, Zweifel BS, Seibert K, Needleman P. Selective regulation of cellular cyclooxygenase by dexamethasone and endotoxin in mice. J Clin Invest 1990;86:1375-79.
2. Wong WYL, Richards J-AS. Evidence for two antigenically distinct molecular weight variants of prostaglandin H synthase in the rat ovary. Mol Endocrinology 1991;5:1269-79.
3. Rosen GD, Birkenmeier TM, Raz A, Holtzman MJ. Identification of cyclooxygenase-related gene and its potential role in prostaglandin formation. Biochem Biophysical Res Commun 1989:164:1358-65.
4. Herschman HR. Primary response genes induced by growth factors and tumor promoters. In: Richardson CC, Abelson JN, Meister A, Walsh CT, editors. Annu Rev Biochem, vol. 60 Palo Alto: Annual Reviews, 1991:281-319.
5. Kujubu DA, Fletcher BS, Varnum BC, Lim RW, Herschman HR. TIS10, a phorbol ester tumor

promoter-inducible mRNA from Swiss 3T3 cells, encodes a novel prostaglandin synthase/cyclooxygenase homologue. J Biol Chem 1991;266:12866-72.

6. Fletcher BS, Kujubu DA, Perrin DM, Herschman HR. Structure of the mitogen-inducible TIS10 gene and demonstration that the TIS10 encoded protein is a functional prostaglandin G/H synthase. J Biol Chem 1992;267:4338-44.
7. Xie W, Chipman JG, Robertson DL, Erikson RL, Simmons DL. Expression of a mitogen-responsive gene encoding prostaglandin synthase is regulated by mRNA splicing. Proc Natl Acad Sci 1991;88:2692-96.
8. O'Banion MK, Sadowski HB, Winn V, Young DA. A serum-and glucocorticoid-regulated 4-kilobase mRNA encodes a cyclooxygenase-related protein. J Biol Chem 1991;266:23261-67.
9. O'Banion MK, Winn VD, Young DA. cDNA cloning and functional activity of a glucocorticoid-regulated inflammatory cyclooxygenase. Proc Natl Acad Sci 1992;89:4888-92.
10. Xie W, Herschman HR. v-src induces Prostaglandin Synthase 2 gene expression by activation of the c-Jun N-terminal kinase and the c-Jun transcription factor. J Biol Chem 1995;270:27622-28.
11. Xie W, Herschman HR. Transcriptional regulation of prostaglandin synthase 2 gene expression by platelet derived growth factor and serum. J Biol Chem 1996;271:31742-48.
12. Xie W, Fletcher BS, Andersen RD, Herschman HR. v-src induction of the TIS10/PGS2 prostaglandin synthase gene is mediated by an ATF/CRE transcription response element. Mol Cell Biol 1994;14:6531-39.
13. Wadleigh DJ, Herschman HR. Transcriptional regulation of the cyclooxygenase-2 gene by diverse ligands in murine osteoblasts. Biochem Biophys Res Commun 1999;264:865-70.
14. Kawata R, Reddy ST, Wolner B, Herschman HR. Prostaglandin synthase 1 and prostaglandin synthase 2 both participate in activation-induced PGD_2 production in mast cells. J Immunol 1995;155:818-25.
15. Reddy ST, Wadleigh DJ, Herschman HR. Transcriptional regulation of the cyclooxygenase-2 gene in activated mast cells. J Biol Chem 2000;275:3107-13.
16. Wadleigh DJ, Reddy ST, Kopp E, Ghosh S, Herschman HR. Transcriptional activation of the cyclooxygenase-2 gene in endotoxin-treated RAW 264.7 macrophages. J Biol Chem 2000; 275:6259-66.
17. Kopp E, Medzhitov R, Carothers J, et al. ECSIT is an evolutionarily conserved intermediate in the Toll/IL-1 signal transduction pathway. Genes Devel 1999;13:2059-71.

ANTIMICROINFLAMMATORY LIPID SIGNALS FROM VASCULAR CYCLOOXYGENASE-2: A NOVEL MECHANISM FOR NSAID AND N-3 PUFA THERAPEUTIC ACTIONS

Charles N. Serhan, Clary B. Clish, Jessica Brannon, Sean P. Colgan, Karsten Gronert, and Nan Chiang

Introduction

Many studies carried out in the past 25 years suggest that dietary omega-3 polyunsaturated fatty acids (ω-3 PUFA) have beneficial effects in human diseases [1,2]. These include antithrombotic, immunoregulatory, and anti-inflammatory responses relevant in arteriosclerosis, arthritis, and asthma [1] as well as antitumor and antimetastatic effects [3]. Their potential for preventative actions in cardiovascular diseases was recently bolstered with the finding that a major dietary ω-3 PUFA, eicosapentaenoic acid (C20:5 ω-3, EPA), has a dramatic effect on ischemia-induced ventricular fibrillation and protects against sudden cardiac death [4]. Due to emergence of such therapeutic actions of ω-3 PUFA, an international workshop [5] has recently called for recommended dietary intakes. However, the molecular mechanism(s) for dietary ω-3 protective actions remain largely unexplained. Many explanations offered [1-5] are not embraced because of the lack of molecular evidence *in vivo* and the high concentrations of ω-3 PUFA required to evoke putative "beneficial actions" *in vitro.* Although the proinflammatory roles of leukotrienes (LT) and prostaglandin (PG) are well appreciated [6], there is new evidence that other eicosanoids derived from arachidonate, namely lipoxins (LXs) and their endogenous analogs, the aspirin-triggered 15 epimer lipoxins (ATL), are potent counterregulators of PMN-mediated injury and acute inflammation [7-9]. Acetylation of cyclooxygenase 2 (COX-2) by aspirin (acetylsalicyclic acid, ASA) prevents the formation of prostanoids, but the acetylated enzyme remains active *in situ*, generating 15-*R*-hydroxyeicosatetraenoic acid (15-*R*-HETE), which is converted by inflammatory cells to 15-epimeric lipoxins (ATL) [10]. Synthetic analogs of these natural local mediators with prolonged bio-half-life display potent anti-inflammatory properties [9,11], providing evidence that cell-cell "cross-talk" can convert arachidonic acid to mediators with anti-inflammatory properties [12].

PMN-vessel interactions are pivotal to recruitment and PMN-dependent tissue injury. Hence, the local signals involved in their "cross-talk dialog" are of interest. Our finding that aspirin-acetylated COX-2 remains active *in vivo* [10] to generate specific ATL that can be effectors of well-established anti-inflammatory therapy offers a mechanism for ASA-beneficial impact that cannot be attributed to prostanoids. New therapeutic uses for

B. Samuelsson et al. (eds.), Advances in Prostaglandin and Leukotriene Research, 49–55.

ASA and related nonsteroidal anti-inflammatory drugs (NSAIDs) continue to be uncovered that require molecular definition, including prophylaxis against colorectal cancer and lowering the risk of myocardial infarction [reviewed in 13], and in view of the apparent overlapping beneficial profiles attributed to dietary ω-3 PUFA in human disease [1-5], we sought evidence for novel mechanisms. Our recent findings [14] reviewed in this contribution uncover the generation of lipid-derived signals that could provide a basis to explain some of the beneficial actions of ω-3 PUFA.

Inflammatory Exudates

We tested the concept that NSAIDs might enable the formation of novel mediators from ω-3 PUFAs. Inflammatory exudates formed in murine air pouches via intrapouch injections of TNFα with ω-3 and ASA on board (2 hours) generated several novel products. Liquid chromotography tandem mass spectrometry (LC/MS/MS) analyses of these exudate-derived materials demonstrated monohydroxy acids, i.e. 18-hydroxy-eicosapentaenoic acid (18-HEPE) and 5-HEPE, which coeluted with synthetic 5*S*-HEPE, as well as novel trihydroxy-containing products derived from C20:5. LC retention times and MS/MS spectra gave product ions consistent with structures shown in the respective insets, namely m/z 317 = $[M-H]^-$, 299 = $[M-H]^- -H_2O$, 273 = $[M-H]^- -CO_2$, 255 = $[M-H]^- -H_2O,-CO_2$. Diagnostic ions consistent with 18-HEPE identification were present at m/z 259 [14] and 5-HEPE at m/z 115. These criteria were used throughout for identification. The chirality at carbon 18 was established for the exudate-derived 18-HEPE using a chiral column and a reference 18*R*-HEPE prepared via biogenic synthesis with *B. megaterium* [14]. The alcohol chirality at position 18 proved to be > 98% *R*. These findings indicated that murine inflammatory exudates exposed *in vivo* to ω-3 20:5 and ASA produced 5-lipoxygenase pathway 5-series 5*S*-HEPE, a product also identified with human PMN [15], as well as the novel 18*R*-HEPE, whose route of formation was determined (Figure 1).

Air pouch inflammatory exudate cells from these ASA- and EPA-treated mice contained predominantly PMN, which were 25-60% lower in number than in exudates formed with TNFα alone (n=3). When activated with ionophore A_{23187}, these exudates generated essentially equivalent amounts of 18*R*-HEPE (10.2 ± 4.3 ng/10^6 cells) and 5*S*-HEPE (10.9 ± 2.9 ng/10^6 cells). Also, several novel trihydroxy-containing products were obtained in these inflammatory exudates. Ions present within MS/MS were consistent with a trihydroxy-containing product from C20:5 with a parent ion m/z 349 = $[M-H]^-$ and product ions of structural significance present at m/z 291 and 195. Also, a 270 nm UV absorbance maximum was evident, indicative of a conjugated triene.

Human Endothelial Cells and Recombinant COX-2

These new products were also generated by human cells. To this end, human endothelial cells known to induce COX-2 with IL-1β [7] or hypoxia were pulsed with EPA, treated with ASA, and extracted materials were subject to LC/MS/MS analyses. Selected ion monitoring at m/z 259 revealed that HUVEC treated with ASA converted EPA to 18*R*-

HEPE. Also, HMVEC treated with ASA and EPA generated 18-HEPE (10.6 ng/10^6 cells) and 15-HEPE (6.0 ng/10^6 cells) (n=2, d=4). These observations implicated COX-2, which proved to be the case.

Recombinant human COX-2 converted linoleic acid (C18:2) to both 13-HODE (ω-5 oxygenation) and 9-HODE (ω-9), which were greatly diminished by ASA but not completely abolished. Arachidonic acid was converted to 15*R*-HETE (ω-5) as well as 11*R*-HETE (ω-9), consistent with earlier findings [7] and with ASA triggered the appearance of a lipoxygenase activity that switched to 15*R*-HETE production by acetylated COX-2, which did not appear to influence 11*R*-HETE formation. 11*R*-HEPE was the major product with EPA and COX-2, with lesser amounts of 15*R*-HEPE (ω-5) and 18*R*-HEPE (ω-2). ASA acetylation of COX-2 led to ~ 2-fold increase in 18*R*-HEPE (ω-2) with a > 85% reduction in 11*R*-HEPE (the ratio of positional oxygenations with C20:5 was 1:1:0.3 with 18*R* ≈ 15*R* > 11*R*). Hence, these findings suggested that acetylated COX-2 in endothelial cells was a dominant source of 18*R*-HEPE and 15*R*-HEPE. Of interest, unlike the isolated COX-2 product profiles, neither 11*R*-HEPE (from C20:5) nor 11*R*-HETE (from C20:4) were major products of the vascular endothelial cells. These results suggest that ASA treatment at local sites of inflammation along with ω-3 PUFA (i.e. EPA; 20:5) administration, as exemplified by cytokine-driven acute inflammation, can convert EPA via induced COX-2 to 18*R*-HEPE and 15*R*-HEPE (Figure 1).

Human Leukocytes

PMN convert ASA-triggered COX-2-derived 15*R*-HETE to 15-epi-LXA_4 (5*S*,6*R*,15*S*-trihydroxy-7,9,13-*trans*-11-cis-eicosatetraenoic acid) and endogenous EPA is converted to 5-series LX by human leukocytes as well as trout macrophages [7]. We questioned whether human PMN engaged in phagocytosis handle acetylated COX-2-derived 20:5 products 18*R*-HEPE and 15*R*-HEPE. Serum-treated zymosan (STZ), the phagocytic stimulus, initiated the utilization and conversion of acetylated COX-2 C20:5-derived products to two classes of trihydroxy-containing EPE, evident again by selected ion monitoring at m/z 349.5 $[M-H]^-$, the expected base peak molecular ion for these products [14]. One gave essentially the same MS/MS observed from murine exudate cells and was consistent with the 5,12,18*R*-triHEPE structure depicted in Figure 1. This product is an 18*R*-hydroxy-carrying "LTB_5-like" structure. Several independent lines of evidence indicated that PMN take up 18*R*-HEPE, which is converted by their 5-lipoxygenase and in subsequent steps to 5-hydroperoxy-18*R*-DiHEPE and 5(6)epoxide formation to 5,12,18*R*-triHEPE (and related isomers), retaining the 18*R* chirality of the precursor. Using a similar biosynthetic route, COX-2-derived 15*R*-HEPE was converted by PMN via 5-lipoxygenation to a 5-series LXA_5 analogue (Figure 1) that also retains its C15 configuration. Its MS/MS gave prominent ions at m/z 305, 233, and 251, namely 15-epi-LXA_5, consistent with LX_5 structure (5 series) observed from endogenous sources of EPA in trout macrophages. In the case of COX-2-derived products generated with NSAIDs, the chirality of the precursor 15*R* is retained by human PMN to give 15-epi-LXA_5, which is the 5-series ω-3 analog of 15-epi-LXA_4. Hence, these results indicate that human endothelial cells and PMN generate the novel products observed with inflammatory exudates.

Antimicroinflammatory Actions of 18*R*- and 15*R*-Derived Products

Transendothelial migration is a pivotal event in PMN recruitment and inflammation; therefore, we isolated and assessed 5,12,18*R*-triHEPE and its precursor 18*R*-HEPE. Both compounds inhibited LTB_4-stimulated PMN transendothelial migration with an apparent IC_{50} at ≈ 50 nM for 5,12,18*R*-triHEPE and IC_{50} ≈ 1.0 μM with 18*R*-HEPE. Thus, the new 5-series members 18*R*-carrying trihydroxy and 18*R*-HEPE inhibited PMN migration, as did 15-epi-LXA_4 and its omega end analog [8,11]. The G protein-coupled surface receptor for LTB_4 was identified, and thus to determine whether these 18*R*-containing products interact with the human BLT receptors to block PMN, BLT was cloned from reported sequences [9] and stably expressed in HEK-293 cells for competition binding experiments. The homoligand LTB_4 effectively competed IC_{50} ≈ 2.5 nM. 18*R*-HEPE did not, while both LTB_5 and 5,12,18*R*-triHEPE competed (IC_{50} ~ 0.5 μM), with a trend for LTB_5 > 5,12,18*R*-triHEPE. Although 5,12,18*R*-triHEPE and a related structure, LTB_5, were substantially less effective than LTB_4, consistent with the reduced LTB_5 PMN activity [15], their potency for displacing [^{3}H]-LTB_4 was in the range of currently available synthetic LTB_4 receptor antagonists (not shown), suggesting that 5,12,18*R*-triHEPE might serve as a damper for BLT-mediated responses *in vivo* if generated in appropriate quantities within the microenvironment as well as a biotemplate for total synthesis of new classes of receptor antagonists. Also, when administered IV tail at low levels (100 ng), 5,12,18*R*-triHEPE was a potent inhibitor of PMN infiltration into murine dorsal air pouches, as was a 15-epi-LX stable analog [11] administered here at equivalent doses for the purpose of direct comparison. 18*R*-HEPE also carried some activity *in vivo* (< 5,12,18*R*-triHEPE).

Common NSAIDs

Other widely used NSAIDs (i.e. acetaminophen and indomethacin) were tested with human recombinant COX-2 and 20:5 (10 μM, 37°C) to determine whether they altered conversion to HEPE. Each inhibited 11-HEPE formation by > 95%. Of interest, 18*R*-HEPE and 15*R*-HEPE formation persisted (~ 1:1 ratio) in the presence of either acetaminophen or indomethacin at concentrations as high as 2 mM, albeit the levels of 15*R*- and 18*R*-HEPE were reduced by 3-8 times their levels in the absence of inhibitors (n=3). These findings indicate that the oxygenation of ω-3 fatty acids to *R*-containing monohydroxy- and monohydroperoxy-containing products is not restricted to ASA treatment and arachidonate. Moreover, these commonly used NSAIDs may still permit PUFA oxygenation by endothelial cells (Figure 1).

Summary

Despite many reports of ω-3 PUFA's (i.e. C20:5) beneficial impact in humans [1-5], oxygenation by COX-2 to generate monohydroxy products has not been addressed in humans or isolated cells. In fish, C20:5 as well as C20:4 are mobilized and utilized in macrophages and platelets to produce both 5-series-derived eicosanoids and 4-series including PG, LT, and LX with ≈ equal abundance. This is not the situation in humans, where C20:4 appears to be preferentially mobilized by phospholipase A_2 for conversion to 4-series eicosanoids [6].

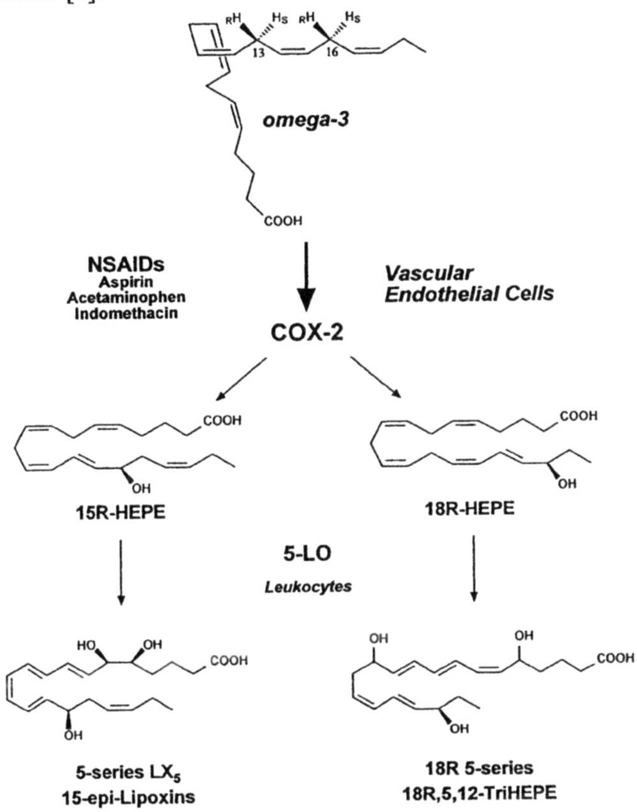

Figure 1. Vascular COX-2 and NSAIDs generate functional arrays of lipid signals from ω-3 PUFA: an example of endogenous inhibitors of microinflammation. Inflammatory cells with upregulated COX-2 treated with NSAIDs oxygenate available ω-3 PUFA converted via a COX-2-NSAID lipoxygenase-type mechanism with stereospecific hydrogen abstraction at C16 or C13 in C20:5 to give *R* insertions of molecular O_2 (e.g. 15*R*-HEPE or 18*R*-HEPE) products that signal with cells within the local microenvironment, reducing recruitment (see text for details). The complete stereochemistry of the novel trihydroxy products remains to be determined. The 1,4-cis-pentadiene structures present in each ω-3 PUFA (i.e., C20:5, C20:3, C22:6) are subject to oxygenation by vascular endothelial COX-2 in a stereospecific fashion. Even in the presence of NSAIDs these *R* monohydroxy products are formed and impact cells relevant in microinflammation and are converted to potent bioactive products by leukocytes.

Inflammatory exudates from mice treated with ASA and EPA generate novel products (Figure 1) that are also produced by human endothelial cells and PMN [14]. Given the 3-5 g amounts of ω-3 PUFA taken as supplements [1,2] and the large area of the microvasculature in humans lined by endothelial cells, the conversion from EPA by these and neighboring cells as observed in our experiments is likely to represent significant amounts at local microenvironments, especially in inflamed or diseased tissues where COX-2 will be focused and a determinant that impacts metabolism [10] when NSAIDs might be of therapeutic benefit, namely microinflammation. Analogous to 15-epi-LX generation, EPA and COX-2-derived 15*R*-HEPE is converted by 5-lipoxygenation and 5(6)epoxide formation in leukocytes to 15-epi-LX_5 (Figure 1). The stable analogs of 15-epi-LXA_4, modified at their C15 position through C20 position with bulky groups, resist inactivating enzymes and are more potent *in vivo*, inhibiting PMN traffic as well as formation and actions of key proinflammatory cytokines [8,11,15]. Hence, the 5-series 15-epi-LX could act in a similar fashion in that they possess a Δ17-18 double bond as the only modification.

Along these lines, the novel 18*R* oxygenation led to products that blocked PMN infiltration *in vivo* and transendothelial migration, providing a basis for a novel mechanism of action to assess the clinical utility of NSAIDs and dietary ω-3 supplementation, namely the endogenous generation of functional arrays of new lipid signals. These new lipids that carry information could, by inhibiting key events in microinflammation (e.g. adhesion and transmigration), mediate some of the beneficial actions noted earlier for ω-3 treatments in human trials [1-5]. Moreover, these findings offer a structural basis to circumvent unwanted effects of current anti-inflammatory therapies as well as potential biochemical indices and/or markers of effective dietary n-3 supplementation.

Acknowledgments

This work was supported in part by National Institutes of Health (NIH) grants GM38765 (C.N.S.) and P01-DE13499 (C.N.S. and S.P.C.). We thank Mary Halm Small for assistance in manuscript preparation.

References

1. De Caterina R, Endres S, Kristensen SD, Schmidt EB, editors. n-3 Fatty Acids and Vascular Disease. London: Springer-Verlag, 1993.
2. Lands WEM, editor. Proceedings of the AOCS Short Course on Polyunsaturated Fatty Acids and Eicosanoids. Champaign, IL: American Oil Chemists' Society, 1987.
3. Iigo M, Nakagawa T, Ishikawa C, et al. Inhibitory effects of docosahexaenoic acid on colon carcinoma 26 metastasis to the lung. Br J Cancer 1997;75:650-5.
4. Billman GE, Kang JX, Leaf A. Prevention of sudden cardiac death by dietary pure ω-3 polyunsaturated fatty acids in dogs. Circulation 1999;99:2452-57.
5. Simopoulos AP, Leaf A, Salem N, Jr. Workshop on the essentiality of an recommended dietary intakes for omega-6 and omega-3 fatty acids. J Am Coll Nutr 1999;18:487-89.
6. Marcus AJ. Platelets: their role in hemostasis, thrombosis, and inflammation. In: Gallin

JI, Snyderman R, editors. Inflammation: Basic Principles and Clinical Correlates. Philadelphia: Lippincott Williams & Wilkins, 1999: 77-95.

7. Clària J, Serhan CN. Aspirin triggers previously undescribed bioactive eicosanoids by human endothelial cell-leukocyte interactions. Proc Natl Acad Sci USA 1995;92:9475-79.
8. Serhan CN, Maddox JF, Petasis NA, et al. Design of lipoxin A_4 stable analogs that block transmigration and adhesion of human neutrophils. Biochemistry 1995;34:14609-15.
9. Chiang N, Gronert K, Clish CB, O'Brien JA, Freeman MW, Serhan CN. Leukotriene B_4 receptor transgenic mice reveal novel protective roles for lipoxins and aspirin-triggered lipoxins in reperfusion. J Clin Invest 1999;104:309-16.
10. Chiang N, Takano T, Clish CB, Petasis NA, Tai H-H, Serhan CN. Aspirin-triggered 15-epi-lipoxin A_4 (ATL) generation by human leukocytes and murine peritonitis exudates: development of a specific 15-epi-LXA_4 ELISA. J Pharmacol Exp Ther 1998;287:779-90.
11. Clish CB, O'Brien JA, Gronert K, Stahl GL, Petasis NA, Serhan CN. Local and systemic delivery of a stable aspirin-triggered lipoxin prevents neutrophil recruitment *in vivo*. Proc Natl Acad Sci USA 1999;96:8247-52.
12. Serhan CN, Takano T, Chiang N, Gronert K, Clish CB. Formation of endogenous "antiinflammatory" lipid mediators by transcellular biosynthesis: lipoxins and aspirin-triggered lipoxins inhibit neutrophil recruitment and vascular permeability. Am J Respir Crit Care Med 2000;161:S95-S101.
13. Levy GN. Prostaglandin H synthases, nonsteroidal anti-inflammatory drugs, and colon cancer. FASEB J 1997;11:234-47.
14. Serhan CN, Clish CB, Brannon J, Colgan SP, Chiang N, Gronert K. Novel functional sets of lipid-derived mediators with antiinflammatory actions generated from omega-3 fatty acids via cyclooxygenase 2-nonsteroidal antiinflammatory drugs and transcellular processing. J Exp Med 2000; in press.
15. Lee TH, Menica-Huerta JM, Shih C, Corey EJ, Lewis RA, Austen KF. Characterization and biologic properties of 5,12-dihydroxy derivatives of eicosapentaenoic acid, including leukotriene B5 and the double lipoxygenase product. J Biol Chem 1984;259:2383-89.

SUBSTRATE INTERACTIONS IN THE CYCLOOXYGENASE-1 ACTIVE SITE

Elizabeth D. Thuresson, Michael G. Malkowski, Karen M. Lakkides, C. Jill Rieke, Ying Sun, Byron Wingerd, Reneé J. Micielli, R. Michael Garavito, and William L. Smith

Introduction

Prostaglandin endoperoxide H synthases-1 and -2 (PGHS-1 and -2) catalyze the conversion of arachidonic acid, two molecules of O_2 and two electrons to PGH_2. This is the committed step in the formation of prostaglandins and thromboxane A_2 [1]. Crystallographic studies of enzyme inhibitor complexes have suggested that the cyclooxygenase active sites of PGHSs are hydrophobic channels that protrude into the body of the major globular domain of the enzymes [2]. We have now determined the structure of arachidonic acid (AA) bound within the cyclooxygenase active site of ovine (o)PGHS-1 [3]. AA is bound in an extended L-shaped conformation and makes a total of 49 hydrophobic contacts (i.e. 2.5-4.0 Å) and two hydrophilic contacts with the protein involving a total of 19 different residues (Figures 1,2). Although AA can assume some 10^7 low energy conformations [4], only three of these are catalytically competent [5]. One conformation leads to PGG_2, one leads to 11*R*-HPETE, and a third leads to 15*R*- plus 15*S*-HPETE. Previous mutational studies have established the importance of Arg120 in AA binding to PGHS-1 and -2 [6-8], the role of Tyr385 in abstraction of the 13-pro*S*-hydrogen from AA [9,10] and the importance of Ser530 and Ile523 as determinants of inhibitor specificity [11-13]. We have performed mutational analyses of a number of the residues that line the cyclooxygenase channel to determine their functional importance in AA binding and oxygenation. Substitutions of several cyclooxygenase site residues lead to large increases in 11-HPETE or 15-HPETE formation but with small changes in the K_m for AA. Our results suggest that individually and collectively the hydrophobic residues function primarily to position AA in a specific conformation that optimizes its conversion to PGG_2.

B. Samuelsson et al. (eds.), Advances in Prostaglandin and Leukotriene Research, 57–64.

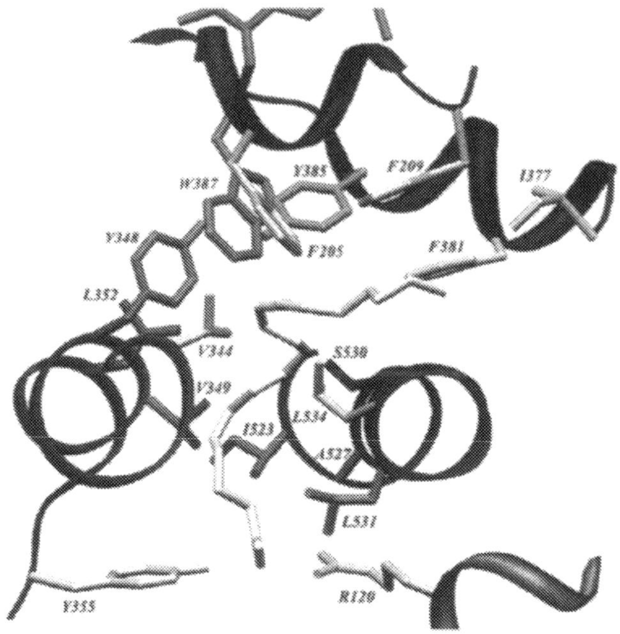

Figure 1. AA bound in the COX channel.

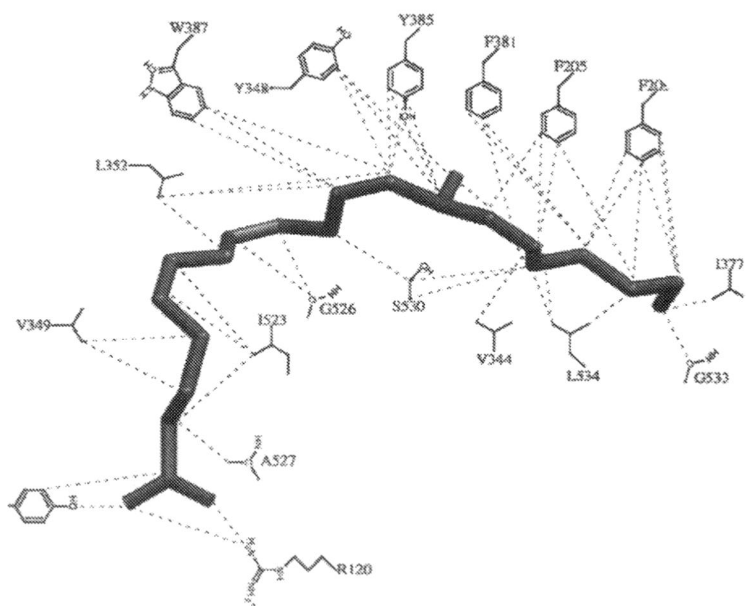

Figure 2. Interactions between AA and COX channel residues. Reprinted with permission from [3].

Mutational Analysis of Cyclooxygenase Active Site Mutants

As illustrated in Figure 1, AA is bound in an extended L-conformation in the AA/Co^{3+}-heme oPGHS-1 co-crystal structure [3]. The carboxylate group of AA interacts with Arg120, the ω end abuts Ile377 and Gly533, the 13-pro*S* hydrogen is appropriately aligned for abstraction by Tyr385, and there is ample space for the first O_2 insertion at C-11 and facile bridging of the incipient 11-hydroperoxyl radical to C-9 to form the endoperoxide. Formation of the cyclopentane ring is proposed to involve rotation about C-11 and consequent movement of the ω terminus so that C-12 can react with the C-8 radical that is produced upon formation of the endoperoxide grouping; this movement, in turn, positions C-15 next to Tyr385 for a second antarafacial O_2 insertion, reduction of the 15-hydroperoxyl radical by Tyr385 and regeneration of the Tyr385 radical [3]. In analyzing various cyclooxygenase active site mutants, we identified the AA oxygenation products, determined kinetic constants for AA oxygenation and measured peroxidase activity (Table 1). The results of these mutational analyses and a discussion of the implications in terms of AA binding and oxygenation are presented below for certain residues.

Residues Critical to High Affinity AA Binding

Arg120. The arginino nitrogen atoms of Arg120 contact both carboxylate oxygens of AA in the AA/Co^{3+}-heme oPGHS-1 co-crystal structure (Figure 1; [3]). Interactions between Arg120 and the carboxylate groups of substrates and inhibitors are important in cyclooxygenase catalysis and inhibition (Table 1; [6,7]). Most notably, replacement of Arg120 with a neutral glutamine causes a 1000-fold increase in the K_m for AA indicating that this residue is essential for high affinity AA binding to the cyclooxygenase site [6,7] and that this involves an ionic linkage between Arg120 and the carboxylate of AA. In contrast, an ionic interaction between Arg120 and AA is not important with PGHS-2 where an R120Q mutation has no detectable effect on either the V_{MAX} or the K_m for AA oxygenation [8].

Residues Directly Involved in Hydrogen Abstraction from C-13 of AA

Tyr385. As reported previously, conversion of Tyr385 to phenylalanine eliminates cyclooxygenase activity without eliminating peroxidase activity [9]. There is considerable evidence that Tyr385 is involved as a tyrosyl radical in abstracting the 13pro*S* hydrogen from AA in the rate determining step in cyclooxygenase catalysis (reviewed in [1]).

Residues Essential for Positioning C-13 of AA for Hydrogen Abstraction

Tyr348. Examination of the crystal structures of oPGHS-1 complexed with nonsteroidal anti-inflammatory drugs [2,14] and the AA/Co^{3+}-heme oPGHS-1 complex (Figure 1; [3]) suggest that there is a hydrogen bond between Tyr348 and Tyr385. Also, the CE2 phenyl ring carbon of Tyr348 is within van der Waals distance of C-12, C-13 and C-14 of AA, suggesting that hydrophobic interactions between Tyr348 and substrate could be important in positioning C-13 of AA with respect to Tyr385. Our results establish that these latter contacts between Tyr348 and AA are essential in positioning C-13 of AA but that there is no functionally important hydrogen bonding between Tyr348 and Tyr385.

Table 1. Kinetic Properties and Product Analyses for oPGHS-1 Cyclooxygenase Active Site Mutants

Enzyme	Cyclooxygenase			[a]Peroxidase (% of Control)	[b]11-HPETE (% of Total Products)	[b]15-HPETE (% of Total Products)
	V_{MAX} (%)	K_m (μM)	V_{MAX}/K_m			
Native	100	2.0	50	100	2.5 (11*R*)	2.5 (15*S*)
R120Q [7]	3	4500	.0007	122	ND	ND
Y348L	0	--	--	9	--	--
Y348F[9,10,15]	83	2.9	29	76	2.1	2.2
V349A	55	1.7	32	52	53 (11*R*)	0
V349S	43	14	3.2	94	41 (11*R*)	0
V349T	39	13	3.0	136	5.5 (11*R*)	0.6
V349L [5]	63	7.1	8.9	117	0	13(15*S*), 11(15*R*)
Y385F [9]	0	--	--	57	--	--
W387A	0	--	--	60	--	--
W387L	7	8.0	0.88	123	53	0.1
W387F [18]	44	25	1.8	57	48	0.2
W387R [18]	0	--	--	0	--	--
W387S [18]	0	--	--	4	--	
G533A [12]	0	--	--	85	--	--
G533A/I523V	0	--	--	54	--	--
L534A	59	8	7.4	72	2	56 (95 *S*, 5*R*)
L534V	98	ND	ND	102	8	47(96*S*, 4*R*)

[a]Peroxidase activity was measured spectrophotometrically using 0.2 mM H_2O_2 and 100 μM TMPD as substrate.

[b]Oxygenase activity was measured by following O_2 consumption with an oxygen electrode. Values are calculated for arachidonic acid turnover by using a value of one mole of arachidonate consumed per two moles of O_2 consumed for native oPGHS-1 and correcting for the percentage of monooxygenated products formed with the mutants.

Substitution of Tyr348 with phenylalanine has little effect on either the cyclooxygenase or peroxidase activity of the enzyme [9,10,15]. The V_{MAX}/K_m of Y348F is 60% of that of native oPGHS-1, and there are no significant differences in product formation between native and Y348F oPGHS-1 (Table 1). Additionally, Y348F oPGHS-1 forms a tyrosyl radical with properties similar to that of native oPGHS-1 [10,15]. These results argue against a significant hydrogen bonding interaction between Tyr348 and Tyr385. Y348L oPGHS-1 lacks cyclooxygenase activity. Examination of the AA/Co^{3+}-heme oPGHS-1 complex (Figure 1; [3]) suggests that substituting leucine at position 348 would permit C-13 of AA to move away from Tyr385, increasing the distance between Tyr385 and C-13 such that hydrogen abstraction could not occur. Thus, we conclude that the phenyl ring of Tyr348 is essential in positioning C-13 of AA with respect to Tyr385.

Gly533. Gly533 lies at the distal end of the cyclooxygenase active site channel with the Cα within van der Waals distance of C-20 of AA (Figure 1, [3]). Substitution of Gly533 with alanine eliminates cyclooxygenase but not peroxidase activity (Table 1; [12]). This contrasts with observations made with murine (m) PGHS-2 where the homologous mutant retains activity toward AA [16]. We prepared and analyzed a G533A/I523V oPGHS-1 mutant to determine if cross-substitution of the one core residue in the cyclooxygenase active site which is different between PGHS-1 and PGHS-2 would permit activity with a G533A mutation in PGHS-1. The G533A/I523V oPGHS-1 retained 54% of the peroxidase activity of native oPGHS-1 but, again, was unable to oxygenate AA at a substrate concentration of 100 μM (Table 1).

RESIDUES CRITICAL FOR CONFORMING AA TO YIELD PGG_2

Val349. One of the methyl groups (CG1) of Val349 lies within van der Waals distance of both C-3 (3.36 Å) and C-4 (3.14 Å) of AA in the AA/oPGHS-1 cocrystal structure (Figure 1; [3]). We substituted this residue with alanine, serine, threonine, and leucine and evaluated these mutants. All of the mutants retained oxygenase and peroxidase activity. The catalytic efficiencies (V_{MAX}/K_m values) for the oxygenase reaction ranged from 6-60% of native oPGHS-1 (Table 1). Mutants in which Val349 was replaced with a smaller residue produced an abundance of 11-HPETE versus PGG_2 and little or no detectable 15-HPETE; the 11-HPETE/PGG_2 ratio for these mutants increased as the size of the group at position 349 was decreased. In contrast, replacing Val349 with a larger leucine residue (V349L oPGHS-1) led to the generation of a relative abundance of 15-HPETE and no detectable 11-HPETE [5].

Chiral HPLC analysis of 11-HPETEs formed from arachidonic acid by V349A, V349S and V349T oPGHS-1 mutants indicated that the product was exclusively 11*R*-HPETE. As reported previously, the V349L oPGHS-1 mutant formed a 65/35 mixture of 15*S*/*R*-HPETE, the same enantiomeric profile obtained with native oPGHS-1 [5].

We reported evidence earlier that AA can assume at least three catalytically competent arrangements in the cyclooxygenase activity site of oPGHS-1 [5]. These arrangements, occurring at the time of hydrogen abstraction, lead to different products--

PGG_2, 11*R*-HPETE or 15*R/S*-HPETE. With native oPGHS-1 the kinetically most favorable arrangement of AA is that which yields PGG_2. With V349A oPGHS-1 the arrangements yielding PGG_2 and 11-HPETE appear to be equally favorable. 11-HPETE would be expected to be formed from an arrangement of AA in which C-9 and C-11 are misoriented such that the endoperoxide bridge between these carbons cannot be formed. The V_{MAX}/K_m ratios are similar for native and V349A oPGHS-1 (Table I). These observations suggest that the valine to alanine substitution does not appreciably influence the positioning of C-13 with respect to Tyr385; moreover, formation of both PGG_2 and 11-HPETE presumably proceed via abstraction of the 13pro*S* hydrogen and formation of an 11-hydroperoxyl radical [17]. Accordingly, we suggest that Val349 plays a major role in positioning the carboxyl half of AA without appreciably affecting the location of the ω half of AA within the cyclooxygenase site. The effect of Val349 on the position of C-9 must occur indirectly because Val349 contacts only C-3 and C-4 of the substrate (Figures 1,2). Placing smaller amino acids at position 349 apparently permits the carboxyl half of AA greater flexibility, which translates into a small shift in the orientation of C-9 with respect to C-11.

Trp387. As shown in Figure 1, Trp387 resides near the apex of the cyclooxygenase channel and is positioned with its CH2 indole ring carbon 3.38 Å from C-11 and 3.65 Å from C-12 of AA; additionally, the CZ2 phenyl ring carbon of Trp387 is 3.96 Å from C-11 of AA [3]. Trp387 also makes a van der Waals contact with Tyr385. Previous studies of W387R, W387F, and W387S oPGHS-1 had indicated that W387R and W387S oPGHS-1 lack cyclooxygenase activity; W387S oPGHS-1 had about 5% of the peroxidase activity of native enzyme (Table 1; [18]). In contrast, W387F exhibits both cyclooxygenase and peroxidase activities and generates a tyrosyl radical signal upon incubation with hydroperoxide [18]. We prepared W387A and W387L oPGHS-1, and performed kinetic studies and product analyses with the W387F mutant. W387A oPGHS-1 lacked cyclooxygenase activity but retained substantial peroxidase activity (Table I). W387L oPGHS-1 had about 10% of the cyclooxygenase activity of native enzyme. The V_{MAX} value for W387F oPGHS-1 was about 45% of that observed with native enzyme although there was about a 10-fold increase in the overall K_m. W387F and W387L oPGHS-1 formed relatively large amounts of 11-HPETE (> 35% of total products) from AA (Table 1); as expected [19], the 11-HPETE was found by chiral HPLC analyses to be exclusively 11*R*-HPETE. As was found for native oPGHS-1 and V349L oPGHS-1 [5], the K_m values for the formation of PGG_2, 11-HPETE, and 15-HPETE were different from one another.

The results of the product analyses of W387F and W387L oPGHS-1 are similar to those observed with V349A oPGHS-1 in that the mutants form large amounts of 11-HPETE. Again, the likely explanation is that the arrangement of AA which leads to 11-HPETE is relatively stabilized with the orientations of C-11 and C-9 militating against endoperoxide formation. In the case of the W387F and W387L oPGHS-1 mutants, there is additional space at position 387. This, in turn, allows AA additional flexibility about C-11 where there are normally van der Waals interactions between this carbon and the CH2 and CZ2 carbons of Trp387. As a consequence AA can more readily assume a conformation that leads to 11-HPETE.

Leu534. The CD1 and CD2 methyl groups of Leu534 are within van der Waals distance of C-15 (3.54 Å), C-16 (3.5 Å) and C-18 (3.91 Å) of AA in the AA/oPGHS-1 crystal structure [3]. Substitution of Leu534 with alanine or valine yields mutant enzymes which produce large amounts of 15-HPETE (Table 1); the 15-HPETE is ≥ 95% 15*S*-HPETE. With native oPGHS-1, the 15*S*-HPETE comprises 65% of the total 15-HPETE. Apparently, having a hydrophobic residue smaller than leucine at position 534 enlarges the cyclooxygenase active site near C-15 providing for relatively greater access of O_2 antarafacial to the site of hydrogen abstraction. 15*S*-HPETE is the only 15-HPETE formed by native hPGHS-2 [19], and the cyclooxygenase site of PGHS-2 is somewhat larger and more accommodating than that of PGHS-1 [1].

All Other Active Site Residues

Other active site residues make lesser but measurable contributions to optimizing catalytic efficiency. Substituting Phe205, Phe209, Tyr355, Phe381, and Ile523 with alanines and Ile377 with valine results in relatively small increases in 11-HPETE formation, indicating that these residues play a subtle role in orienting C-11 with respect to C-9 at the time of endoperoxide formation, either by helping to align the carboxyl half of AA (Tyr355, Ile523) or by stabilizing the methyl end of AA. With the exception of I377V, these same mutant enzymes also cause a modest decrease in the V_{max} of the enzyme, indicating that they also play a role in positioning C-13 with respect to Tyr385. Ser353, when substituted with threonine, also causes a slight change in product distribution, resulting in an overproduction of 11- and 15-HPETE, indicating that this residue somehow helps to preserve proper orientation of C-9 with respect to C-11, while at the same time, promoting efficient oxygenation of C-11.

Replacement of Ser530, the site of aspirin acetylation [11,14], with threonine, causes a 95% decrease in catalytic efficiency and an increase in only 15*R*-HPETE, but not 15*S*-HPETE; S530V is catalytically inactive. Although acetylation of oPGHS-1 eliminates cyclooxygenase activity, aspirin-acetylated PGHS-2 forms exclusively 15*R*-HPETE [19,20] and involves removal of the 13-pro*S* hydrogen [17]. This would imply that 15*R*-HPETE formation involves suprafacial addition of O_2 to C-15. Addition of a residue slightly larger than serine at position 530 might be expected to block antarafacial O_2 addition without preventing abstraction of the 13-pro*S* hydrogen. Replacement with the larger valine would narrow the cyclooxygenase channel so that hydrogen abstraction could not occur.

References

1. Smith WL, DeWitt DL, Garavito RM. Cyclooxygenases: structural, cellular and molecular biology. Annu Rev Biochem 2000;69:149-82.
2. Picot D, Loll PJ, Garavito M. The X-ray crystal structure of the membrane protein prostaglandin H2 synthase-1. Nature 1994;367:243-49.
3. Malkowski MG, Ginell S, Smith WL, Garavito RM. The x-ray crystal structure of prostaglandin endoperoxide H synthase-1 complexed with arachidonic acid. Science 2000; in press.
4. Gund P, Shen TY. A model for the prostaglandin synthetase cyclooxygenation site and its inhibition by antiinflammatory arylacetic acids. J Med Chem 1977;9:1146-52.

5. Thuresson ED, Lakkides KM, Smith WL. Different catalytically competent arrangements of arachidonic acid within the cyclooxygenase active site of prostaglandin endoperoxide H synthase-1 lead to the formation of different oxygenated products. J Biol Chem 2000; 275:8501-7.
6. Mancini JA, Riendeau D, Falgueyret JP, Vickers PJ, O'Neill GP. Arginine 120 of prostaglandin G/H synthase-1 is required for the inhibition by nonsteroidal anti-inflammatory drugs containing a carboxylic acid moiety. J Biol Chem 1995;270(49): 29372-77.
7. Bhattacharyya DK, Lecomte M, Rieke CJ, Garavito M, Smith WL. Involvement of arginine 120, glutamate 524, and tyrosine 355 in the binding of arachidonate and 2-phenylpropionic acid inhibitors to the cyclooxygenase active site of ovine prostaglandin endoperoxide H synthase-1. J BiolChem 1996;271(4):2179-84.
8. Rieke CJ, Mulichak AM, Garavito RM, Smith WL. The role of arginine 120 of human prostaglandin endoperoxide H synthase-2 in the interaction with fatty acid substrates and inhibitors. J Biol Chem 1999;274(24):17109-14.
9. Shimokawa T, Kulmacz RJ, DeWitt DL, Smith WL. Tyrosine 385 of prostaglandin endoperoxide synthase is required for cyclooxygenase catalysis. J Biol Chem 1990; 265(33):20073-76.
10. Tsai A, Hsi LC, Kulmacz RJ, Palmer G, Smith WL. Characterization of the tyrosyl radicals in ovine prostaglandin H synthase-1 by isotope replacement and site-directed mutagenesis. J Biol Chem 1994;269(7):5085-91.
11. DeWitt DL, El-Harith EA, Kraemer SA, et al. The aspirin and heme-binding sites of ovine and murine prostaglandin endoperoxide synthases. J Biol Chem 1990;265(9):5192-98.
12. Shimokawa T, Smith WL. Prostaglandin endoperoxide synthase: the aspirin acetylation region. J Biol Chem 1992;267:12387-92.
13. Guo Q, Wang LH, Ruan KH, Kulmacz RJ. Role of Val509 in time-dependent inhibition of human prostaglandin H synthase-2 cyclooxygenase activity by isoform-selective agents. J Biol Chem 1996;271(32):19134-39.
14. Loll PJ, Picot D, Garavito RM. The structural basis of aspirin activity inferred from the crystal structure of inactivated prostaglandin H2 synthase. Nat Struct Biol 1995;2:637-643.
15. Hsi LC, Hoganson CW, Babcock GT, Garavito RM, Smith WL. An examination of the source of the tyrosyl radical in ovine prostaglandin endoperoxide synthase-1. Biochem Biophys Res Commun 1995;207(2):652-60.
16. Rowlinson SW, Crews BC, Lanzo CA, Marnett LJ. The binding of arachidonic acid in the cyclo-oxygenase site of mouse prostaglandin endoperoxide sythase-2 (COX-2). J Biol Chem 1999;274(33):23305-10.
17. Schneider C, Brash AR. Stereospecificity of hydrogen abstraction in the conversion of arachidonic acid to 15R-HETE by aspirin-treated cyclooxygenase-2: implications for the alignment of substrate in the active site. J Biol Chem 2000;275:4743-46.
18. Hsi LC, Tsai AL, Kulmacz RJ, et al. Trp387 and the putative leucine zippers of PGH synthases-1 and -2. J. Lipid Mediat 1993;6(1-3):131-38.
19. Xiao G, Tsai AL, Palmer G, Boyar WC, Marshall PJ, Kulmacz RJ. Analysis of hydroperoxide-induced tyrosyl radicals and lipoxygenase activity in aspirin-treated human prostaglandin H synthase-2. Biochemistry 1997;36(7):1836-45.
20. Lecomte M, Laneuville O, Ji C, DeWitt DL, Smith WL. Acetylation of human prostaglandin endoperoxide synthase-2 (cyclooxygenase-2) by aspirin. J Biol Chem 1994; 269(18):13207-15.

MUTATION OF THE HUMAN PROSTACYCLIN SYNTHASE GENE

Tomohiro Nakayama, Masayoshi Soma, Duolikun. Rehemudula, and Katsuo Kanmatsuse

Introduction

Biosynthesis of prostacyclin (PGI_2) is catalyzed serially by phospholipase A_2 which liberates arachidonic acid from the prostaglandin H_2 synthase, a bifunctional enzyme which catalyzes the conversion of arachidonic acid to prostaglandin G_2 and subsequently to prostaglandin H_2, and PGI_2 synthase (PGIS, EC 5.3.99.4) which catalyzes the formation of PGI_2 from prostaglandin H_2. PGIS has been shown to be widely distributed, predominantly in vascular endothelial and smooth muscle cells [1,2]. An impaired synthesis of PGI_2 has been implicated in the development of essential hypertension (EH), or cerebral infarction (CI). We already determined the organization of this gene [3]. The purpose of this study was to search for possible point mutations in all 10 exons and 5'-flanking region of the PGIS gene in patients with EH or CI.

Materials and Methods

SEARCHING THE MUTATION OF PGIS GENE

Single-strand conformation polymorphism (SSCP) analysis was performed as described previously [4]. DNA was extracted from whole blood according to standard procedures. Polymerase chain reaction (PCR)-SSCP was performed on DNAs from 90 patients with CI.

DNAs from a patient presenting variant electrophoretic patterns and from controls were reamplified by PCR and direct sequencing was performed.

SUBJECTS FOR ASSOCIATION STUDY

The association study between EH and normotensives consisted of 125 patients (mean age 53.0 ± 6.8 years) with EH diagnosed as described previously [5].

The association study between CI and non-CI consisted of 111 patients (mean age 65.8 ± 9.6 years) with CI diagnosed by computed tomography (CT) or magnetic resonance imaging (MRI). All patients had neurological deficits that persisted for at least 1 month. One hundred fifty-two subjects without CI (mean age 66.5 ± 4.6 years) were studied as

B. Samuelsson et al. (eds.), Advances in Prostaglandin and Leukotriene Research, 65–68.

controls. Control subjects had vascular risk factors such as hypertension, hypercholesterolemia, or diabetes mellitus, but no cerebrovascular disease. Individuals with atrial fibrillation were excluded from both the CI and non-CI groups. Informed consent was obtained from each subject according to a protocol approved by the Human Studies Committee of Nihon University.

Transcriptional Activity

To make human PGIS reporter gene constructs, the PCR product was reamplified and subcloned into the TA cloning vector for PCR products (Invitrogen, Carlsbad, CA). After digestion with *Sac* I and *Xho* I, each product was subcloned into the *Sac* I/*Xho* I sites of the luciferase reporter gene vector, pGV-B2 (Toyo Inkt Inc., Tokyo, Japan). All constructs were verified by sequencing the inserts and flanking regions of the plasmids.

Human aortic smooth muscle cells (HASMC) were purchased from Kurabo (Osaka, Japan) and cultured. Transfection was performed as described previously [5].

Statistical Analysis

Allele frequencies were calculated from the genotypes of all subjects. Hardy-Weinberg equilibrium was assessed by χ^2 analysis. The overall distribution of alleles was analyzed by 2 × allele number contingency tables, and a p-value of less than 0.05 was considered significant. Individual differences of allele frequencies were tested using 2 × 2 contingency tables for each allele by combining the remaining alleles into one category, and a p-value of less than 0.01 (i.e. 0.05/5) was considered significant to correct for the number of comparisons made. The association of CI and genotypes was evaluated by multiple logistic regression analysis. CI was regarded as the dependent variable, and the number of the repeats, age, and sex were considered independent variables. The overall distributions of alleles between hypertensive patients and non-hypertensives, and between diabetic patients and non-diabetics were assessed by χ^2 analysis. Differences in transcriptional activities of polymorphic alleles were analyzed with one-way repeated measure analysis, and a p-value of less than 0.05 was considered significant.

Results and Discussion

One subject showed an abnormally migrating band on the analysis of exon 2. Sequencing of the mutant type revealed a nonsense mutation in codon 26. So CGA, the codon of Arginine, is changed to TGA the stop codon. This change of nucleotide makes *Bst E*II the restriction site. By PCR and *Bst E*II digestion, 300 subjects including 150 EH subjects and 150 healthy controls were screened. The mutation was found in one patient with EH and in none of the control group members.

This patient was shown to be heterozygous for this mutation. The patient is a 57-year-old woman who had a blood pressure at 177/113 mmHg and her electrocardiogram revealed left ventricular hypertrophy. Although she had never smoked and rarely consumed alcohol, she had a history of CI at the age of 50.

We analyzed this mutation on the family members of this patient. Her father died of stroke at age of 70. Her mother was healthy until her death at age of 92. She had 8 siblings including 2 elder brothers, one who had died in a war and one who had died in chronic renal failure. Another 5 siblings were analyzed on the mutation, and 3 had this mutation. All subjects with the mutation were hypertensives. One hypertensive with mutation had also the history of a CI. We speculate that her and her brother's history of CI and her father's stroke may have a relation to the abnormality of the PGIS.

Furthermore we found a novel polymorphism varied from 3 to 7 repeats of nine nucleotides of Spl site and AP-2 site in core promoter element. This site may have implications for the activity of promoter.

The transcriptional activity of the promoter region allele was determined by luciferase assay. Luciferase reporter gene constructs were then transiently transfected into human aortic smooth muscle cells for determination of PGIS promoter activity. The transcriptional activity increased with increasing numbers of repeats.

We performed an association study using this polymorphism in 125 EH patients and 125 controls. The allele frequency distribution in the two groups were not significantly different. Thus, this polymorphism in the PGIS gene is not associated with EH.

Next, we performed an association study in 111 patients with CI and 152 non-CI controls. The overall distribution of alleles differed significantly between both groups. The association of CI and genotypes was evaluated by multiple logistic regression analysis. CI was regarded as the dependent variable, and the number of the repeats, age, and sex were considered independent variables. This was significant different (odds ratio 1.38, 95% confidence interval, 1.11 to 1.71). The logistic linear regression curve between CI and the total number of repeats for both chromosomes revealed the small number repeat allele to be found more frequently with CI. In the present study, the clinical data for the risk of CI is in good agreement with the experimental data from reporter gene analysis of transcriptional activity. Our study suggests that prostacyclin function depends on the different alleles of the prostacyclin synthase gene and may influence the risk of CI.

In conclusion: We found a nonsense mutation of PGIS gene in a family and a variable number of tandem repeat polymorphism in the promoter region associated with CI.

Acknowledgments

We wish to acknowledge Ms. H. Tobe for her technical assistance, and Drs. Y. Takahashi, J. Uwabo, M. Satoh, A. Kubo, M. Kunimoto, N. Fukuda, Y. Watanabe, and Y. Izumi for collecting samples. This work was supported financially by a grant from the Ministry of Education, Science and Culture of Japan (High-Tech Research Center, Nihon University, no. 10671044), and the Tanabe Biomedical Conference, Japan.

References

1. Ingerman-Wojenski C, Silver MJ, Smith JB, Macarak E. Bovine endothelial cells in culture produce thromboxane as well as prostacyclin. J Clin Invest 1981;67:1292-96.
2. Smith WL, DeWitt DL, Allen ML. Bimodal distribution of the prostaglandin I_2 synthase antigen in smooth muscle cells. J Biol Chem 1983;258:5922-26.
3. Nakayama T, Soma M, Izumi Y, Kanmatsuse K. Organization of the human prostacyclin synthase gene. Biochem Biophys Res Commun 1996;221:803-6.
4. Nakayama T, Soma M, Rahmutula D, Izumi Y, Kanmatsuse K. Nonsense mutation of prostacyclin synthase gene in a family. Lancet 1997;349:1887-88.
5. Nakayama T, Soma M, Takahashi Y, Rehemudula D, Kanmatsuse K, Furuya K. Functional deletion mutation of the 5'-flanking region of the type A human natriuretic peptide receptor gene and its association with essential hypertension and left ventricular hypertrophy in the Japanese. Circ Res 2000;86:841-45.

STRUCTURE AND FUNCTION OF TWO DISTINCT TYPES OF PROSTAGLANDIN D SYNTHASE

Yoshihiro Urade and Osamu Hayaishi

Introduction

Prostaglandin (PG) D synthase (PGDS, EC 5.3.99.2) catalyzes the isomerization of PGH_2, a common precursor of various prostanoids, to produce PGD_2 in the presence of sulfhydryl compounds. PGD_2 induces sleep [1], regulates nociception, inhibits platelet aggregation, acts as an allergic mediator [2], and is further converted to 9α, 11β-PGF_2 or the J series of PGs, such as PGJ_2, Δ^{12}-PGJ_2, and 15-deoxy-$\Delta^{12,14}$-PGJ_2. 15-Deoxy-$\Delta^{12,14}$-PGJ_2 is a ligand for PPARγ, a nuclear receptor involved in differentiation of adipocytes, macrophages, and monocytes. Two distinct types of PGDS have been identified; one is the lipocalin-type enzyme (L-PGDS); and the other, the hematopoietic enzyme (H-PGDS). For each enzyme, we isolated the cDNA and the gene, determined the X-ray crystallographic structure, tissue distribution profile, and the cellular localization in several animal species, and generated gene-knockout (KO) and human enzyme-overexpressing transgenic (TG) mice. Although L-PGDS and H-PGDS catalyze the same reaction, they are quite different from each other in terms of their catalytic properties, amino acid sequence, tertiary structure, evolutional origin, gene structure, chromosomal localization, cellular localization, tissue distribution, and also functional relevance, as summarized in Table 1. Therefore, we consider that L-PGDS and H-PGDS are a novel example of functional convergence. We have recently reviewed a part of these findings elsewhere [3,4].

Lipocalin-type PGD Synthase

We purified L-PGDS from rat brain as an acidic, monomeric soluble protein with a M_r of 26,000. L-PGDS requires free sulfhydryl compounds, such as β-mercaptoethanol, DTT, or GSH, for the reaction. L-PGDS is now known to be an N-glycosylated protein with a M_r of 20,000 to 31,000 depending on the size of the glycosyl moiety and to be the same protein as β-trace, a major protein of human cerebrospinal fluid.

The cDNA for L-PGDS was isolated from many mammalian species, including human and mouse, and also from two amphibians. A homology search in data bases of protein primary structure revealed that L-PGDS is a member of the lipocalin (lipophilic ligand-carrier protein) gene family, which is composed of a variety of secretory proteins, such as β-lactoglobulin and plasma retinol-binding protein, involved in the binding and

B. Samuelsson et al. (eds.), Advances in Prostaglandin and Leukotriene Research, 69–72.

transport of small hydrophobic molecules. L-PGDS is the first member of the lipocalin family to be recognized as an enzyme and a highly glycosylated protein. L-PGDS also binds retinoids, bilirubin and biliverdin with high affinities (K_d = 30 to 80 nM), similar to other lipocalins. Therefore, we consider that L-PGDS is a dual functional protein; it acts as a PGD_2-producing enzyme within cells and functions as a lipophilic ligand-binding protein after secretion into the extracellular space or into various body fluids. Recombinant mouse L-PGDS was recently crystallized by our group. The X-ray diffraction analysis revealed that L-PGDS possesses a β-barrel structure with a hydrophobic pocket. An active thiol, Cys^{65}, the active center for the catalytic reaction, is located facing to the inside of the pocket.

Table 1. Comparison of L-PGDS and H-PGDS.

	L-PGDS	H-PGDS
M_r	20,000-31,000	26,000
Cofactor	β-Mercaptoethanol, DTT, GSH	GSH
Post-translational modification	Cleavage of a signal peptide N-Glycosylation	
Subcellular distribution	Rough ER & nuclear membrane Extracellular fluid (β-trace)	Cytosol
Other activity	Binding of retinoids and bilirubin	GST
Gene family	Lipocalin	GST (sigma-class)
Tertiary structure (Accession No)	β-Barrel	α+β_(1PD2)
cDNA Rat	(J04488, M61902)	(D82071)
Mouse	(X89222)	(D82072)
Human	(M61901)	(D82073)
Amino acid residues	189- 190	199
Gene Rat	(M94134)	
Mouse	(Y10138, D83329)	(AB008824)
Human	(M98537, M98538)	(AB008825-30)
Chromosomal localization Mouse	2B-C1	3D-E
Human	9q34.2-34.3	4q21-22
Transcriptional activator (Responsive element)	Thyroid hormone (TRE) Glucocorticoid (GRE)	(Oct-1) (AP-2)

The gene for L-PGDS was cloned from rat, human, and mouse sources and shown to span about 3 kb and contain seven exons split by six introns. The gene organization is remarkably analogous to that of other lipocalins in terms of number and size of exons and phase of splicing of introns. The mouse and human genes were mapped to chromosome 2B-C1 and 9q34.2-34.3, respectively, both of which were localized within the lipocalin gene cluster. The gene expression for L-PGDS is upregulated by thyroid hormone or dexamethasone with the transcriptional activator being TRE or GRE, respectively[5].

L-PGDS is localized in the central nervous system and male genital organs of various mammals, as well as in the human heart, and is secreted into cerebrospinal fluid, seminal plasma, and plasma, respectively. L-PGDS expression is upregulated in human vascular endothelial cells by fluid shear stress [6]. The urinary L-PGDS concentration is significantly increased in diabetic patients having a normal plasma creatinine level at the early stage of the disease [7]. Therefore, the L-PGDS concentration may be a useful clinical marker for various diseases.

L-PGDS-KO-mice grow normally but show several functional abnormalities in nociception and sleep. The KO-mice do not exhibit allodynia (touch-evoked pain), a typical phenomenon of neuropathic pain, after an intrathecal administration of PGE_2 or bicuculline, a $GABA_A$-receptor antagonist. The KO-mice also show less rebound sleep after sleep deprivation than the wild-type mice. Alternatively, the human L-PGDS-overexpressing TG-mice showed a remarkable time-dependent increase in nonrapid eye movement sleep after stimulation by tail clipping [8] and increases in allergic reactivity.

Hematopoietic PGD Synthase

H-PGDS was originally purified from rat spleen as a cytosolic, GSH-requiring enzyme with a M_r of 26,000. H-PGDS absolutely requires GSH for the reaction and is associated with the activity of GSH S-transferase (GST; EC 2.5.1.18) at about 10% of the turnover rate of PGDS activity.

The cDNA for H-PGDS was cloned from rats, chickens, mice, and humans and found to encode a protein composed of 199 amino acid residues without a signal peptide or N-glycosylation sites. By a homology search in data bases of protein primary structure, H-PGDS was shown to be a member of the GST family. However, H-PGDS showed a weak homology against mammalian GST of the alpha, mu, pi, and theta classes yet a high homology with invertebrate GST of the sigma class. Since the sigma class GST had been observed only in invertebrates, H-PGDS is the first vertebrate homolog of the sigma class GST. We crystallized rat H-PGDS complexed with GSH and determined its tertiary structure with a resolution of 2.3Å by X-ray diffraction analysis. The overall structure is similar to that of other GST but it possesses a unique catalytic site within a wide cleft near the GSH-binding site. Site-directed mutagenesis study revealed that Lys^{112}, Cys^{156}, and Lys^{198} are involved in the binding of PGH_2, Trp104 is critical for structural integrity of the catalytic center, and Tyr^8 and Arg^{14} are essential for activation of the thiol group of GSH [9].

The human and mouse genes for H-PGDS were cloned [10] and their chromosomal localizations were mapped to 4q21-22 and 3D-E, respectively, which are different from those of any of human and mouse genes for the alpha, mu, pi, and theta classes of the GST family. The human and mouse H-PGDS genes are the largest in size (41 and 28 kb, respectively) among genes for the members of the GST family. The genomic structure of H-PGDS is the most similar to that of the invertebrate sigma GST, suggesting that the genes for H-PGDS and the invertebrate sigma GST evolved from a common ancestor.

H-PGDS is widely distributed in the peripheral tissues and localized in the antigen-presenting cells, mast cells, megakaryocytes, and type 2 helper T lymphocytes [11], although the tissue distribution profiles of the mRNA for H-PGDS are remarkably different among animal species. The gene expression for H-PGDS is upregulated by c-Kit ligand in mouse mast cells and by phorbol ester in human megakaryocytes. The transcriptional activators for the human H-PGDS gene were identified to be Oct-1 and AP-2 [12].

H-PGDS-KO-mice and the human H-PGDS-overexpressing TG-mice also grow normally. The H-PGDS-TG mice showed a pronounced adipogenesis under a high fat diet and developed an enhanced allergic lung inflammation in response to antigen challenge.

References

1. Urade Y, Hayaishi O. Prostaglandin D_2 and sleep regulation. Biochim Biophys Acta 1999; 1436:606-15.
2. Matsuoka T, Hirata M, Tanaka H, et al. Prostaglandin D_2 as a mediator of allergic asthma. Science 2000;287:2013-17.
3. Urade Y, Hayaishi O. Prostaglandin D synthase: structure and function. Vitam Horm 2000;58:89-120.
4. Urade Y, Hayaishi O. Biochemical, structural, genetic, physiological, and pathophysiological features of lipocalin-type prostaglandin D synthase. Biochim Biophys Acta 2000; in press.
5. García-Fernández LF, Iñiguez MA, Eguchi N, Fresno M, Urade Y, Muñoz A. Dexamethasone induces lipocalin-type prostaglandin D synthase gene expression in mouse neuronal cells. J Neurochem 2000;75:460-70.
6. Tabe Y, Sasaguri T, Miyagi M, et al. Fluid shear stress induces lipocalin-type prostaglandin D_2 synthase in vascular endothelial cells. Circ Res 2000;86:967-73
7. Hirawa N, Uehara Y, Ikeda T, et al. Urinary prostaglandin D synthase (β-trace) excretion increases in the early-stage of diabetes mellitus. Nephron 2000; in press.
8. Pinzar E, Kanaoka Y, Inui T, Eguchi N, Urade Y, Hayaishi O. Prostaglandin D synthase gene is involved in the regulation of non-rapid eye movement sleep. Proc Natl Acad Sci USA 2000;97:4903-7.
9. Pinzar E, Miyano M, Kanaoka Y, Urade Y, Hayaishi O. Structural basis of hematopoietic prostaglandin D synthase activity elucidated by site-directed mutagenesis. J Biol Chem 2000;275:31239-44
10. Kanaoka Y, Fujimori K, Kikuno R, Sakaguchi Y, Urade Y, Hayaishi O. Structure and chromosomal localization of human and mouse genes for hematopoietic prostaglandin D synthase. Conservation of the ancestral genomic structure of sigma-class glutathione S-transferase. Eur J Biochem 2000;267:3315-22.
11. Tanaka K, Ogawa K, Sugamura K, Nakamura M, Takano S, Nagata K. Differential production of prostaglandin D_2 by human helper T cell subsets. J Immunol 2000;164: 2277-80.
12. Fujimori K, Kanaoka Y, Sakaguchi Y, Urade Y. Transcriptional activation of human hematopoietic prostaglandin D synthase gene in megakaryoblastic cells. J Biol Chem 2000; in press.

Roles of Prostaglandin F Synthase

Kikuko Watanabe, Toshiko Suzuki-Yamamoto, Mikio Nishizawa, Motonari Fukui, Kazunori Ishimura, and Seiji Ito

Introduction

Prostaglandin (PG) F_2 is widely distributed in various organs, and exhibits various biological actions such as smooth muscle contraction of uterus [1], bronchus [2], and trachea [3], the initiation of parturition [4], and pain transmission [5]. PGF synthase (PGFS) is a dual function enzyme, which catalyzes the reduction of not only PGD_2 but also PGH_2 on the same molecule [6]. PGFS exhibits reductase activities toward various carbonyl compounds [6], such as phenanthrenequinone (PQ) in addition to PGD_2 and PGH_2. PGFS belongs to the aldo-keto reductase (AKR) superfamily based on substrate specificity, molecular weight, and the amino acid sequence [7-10]. Recently, we isolated the clones of four human enzymes [9,10], including PGFS, which belong to the AKR family. Although the amino acid sequences of these enzymes show high homology (83-98%), only PGFS, which was classified to AKR1C3 [10], catalyzed the reduction of PGD_2 and PGH_2, and the oxidation of 9α, 11β-PGF_2 [9, 10]. Moreover, we developed a high-stringency PCR technique to discriminate fine sequence differences using each specific primer, and examined RT-PCR analysis to estimate expression of AKR mRNAs [10]. The tissue distribution of these four enzymes was different, and human PGFS was localized in lung, liver, kidney, muscle, and peripheral blood lymphocyte. The expression in lymphocytes was markedly suppressed by the T cell mitogen concanavalin A. Moreover, PGFS was abundant in the contractile interstitial cells (CIC) in the alveolar septum of bovine lung, and was diffusely present in the cytoplasm of the endothelial cells of bovine liver. These results suggest that PGFS plays important roles in both the physiological and pathophysiological processes.

The Amino Acid Sequence of Human PGFS, Belonging to Aldo-keto Reductase Family, its Substrate Specificity, and Tissue Distribution

We isolated a cDNA clone of PGFS from human lung, and other three human cDNA clones of 20α-hydroxysteroid dehydrogenase (20α-HSD), bile acid binding protein (BABP), and dihydrodiol dehydrogenase (DD) 4, which belong to 1C group of the AKR family. Although the amino acid sequences of these four enzymes showed about 90% identity, only

B. Samuelsson et al. (eds.), Advances in Prostaglandin and Leukotriene Research, 73–76.

a PGFS catalyzed the reduction of PGD_2 and PGH_2, and the oxidation of 9α, 11β-PGF_2, and other enzymes could not reduce PGs. (Figure 1). To distinguish PGFS from other enzymes, we developed a high-stringency PCR technique to discriminate fine sequence difference using each specific primer.

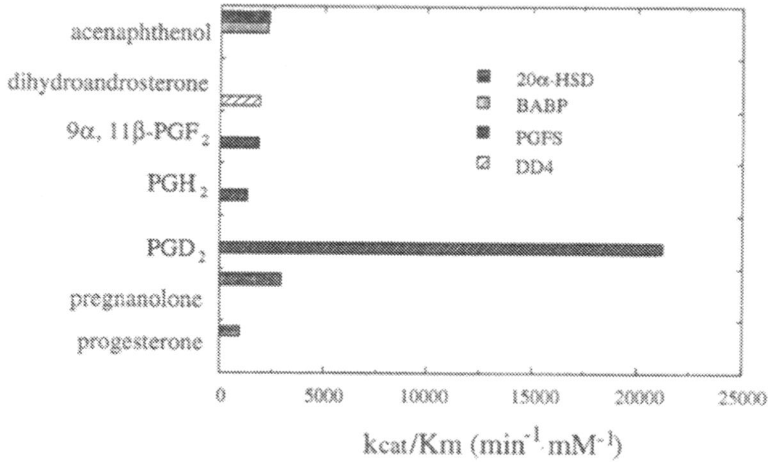

Figure 1. The substrate specificities of 20α-HSD, BABP, PGFS, and DD4.

The clones were denatured at 94°C for 10 minutes, and subjected to step-down PCR with each specific primer and *Gene Taq* DNA polymerase. The step-down protocol applied to the PCR was 10 cycles (94°C, 1 minute; 72°C, 3 minutes), 10 cycles (94°C, 1 minute; 65°C, 2 minutes; 72°C, 1 minute), and 15 cycles (94°C, 1 minute; 60°C; 2 minutes; 72°C, 1 minute). Using this method, the specific size of band of each enzyme clone was detected (three upper columns of Figure 2). However, when the common primer of four enzymes was used, or when a standard PCR protocol was applied using specific primers with thermostable polymerase, all the clones gave the amplification (the bottom column of Figure 2).

Using this high-stringency PCR technique, we examined the mRNA distribution in human tissues. PGFS was expressed ubiquitously in lung, liver, kidney, muscle, peripheral blood lymphocytes (PBL), and uterus. Especially, PGFS was expressed in PBL. As PGD_2 is increased in alveolar lymphocytes as well as in macrophage, eosinophils, and mast cells after a challenge with an inflammatory antigen, we examined the effect of Con A. Con A is considered to be a T lymphocyte-specific mitogen, on expression of PGFS mRNA in PBL. The incubation of PBL for 24 hours with Con A suppressed the expression of PGFS mRNA in PBL. These results suggest that human PGFS plays an important role in pathogenesis of allergic diseases such as asthma.

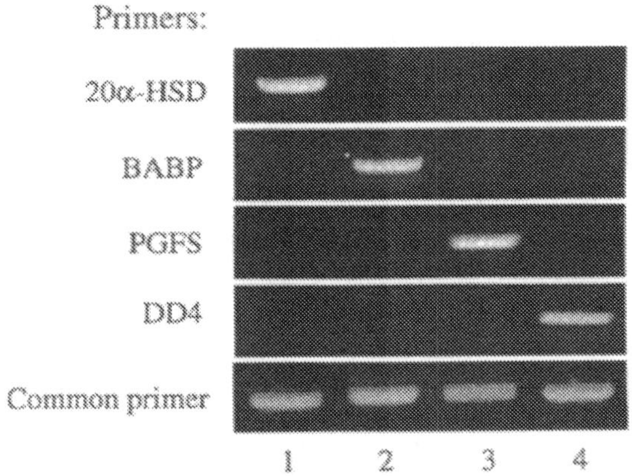

Figure 2. The high-stringency PCR of four enzyme clones of aldo-keto reductase family with specific primers.

The Immunohistochemical Studies of PGFS in Bovine Lung and Liver

Using the anti-bovine lung PGFS antibody and α-smooth muscle actin, which is specific for smooth muscle, the PGFS-positive cells in alveolar septum was identified by immunofluorescence labeling and immunoelectron microscopy [11,12]. PGFS is located in contractile interstitial cells (CIC), but is not located in smooth muscle. CIC is named by Kapanci et al. [13], and is related to the hypoxia-induced contraction. These cells occupied 42% of total interstitial volume of the alveolar walls, and the contraction of CIC changes the configuration of pulmonary capillaries, resulting in augmentation of vascular resistance. Moreover, the results of double immunostaining of PGFS and COX-1 suggest that the sinusoidal endothelial cells have COX I and PGFS, and that they synthesize PGF_2 [14].

References

1. O'Brien WF. The role of prostaglandins in labor and delivery.Clin Perinatol 1995;22(4): 973-84.
2. Panczenko B, Grodzinska L, Gryglewski R. The dual action of meclofenamate on the contractile response to PGF2 alpha in the guinea-pig trachea. Pol J Pharmacol Pharm 1975;27(3):273-76.
3. Dunlop LS, Smith AP. Reduction of antigen-induced contraction of sensitized human bronchus in vitro by indomethacin. Br J Pharmacol 1975;54(4):495-97.
4. Sugimoto Y, Yamasaki A, Segi E, et al. Failure of parturition in mice lacking the prostaglandin F receptor. Science 1997;1;277(5326):681-83.
5. Minami T, Uda R, Horiguchi S, Ito S, Hyodo M, Hayaishi O. Allodynia evoked by intrathecal administration of prostaglandin F2 alpha to conscious mice. Pain 1992;50(2): 223-29.
6. Watanabe K, Yoshida R, Shimizu T, Hayaishi O. Enzymatic formation of prostaglandin F2 alpha from prostaglandin H2 and D2. J Biol Chem 1985;260:7035-41.

7. Watanabe K, Fujii Y, Nakayama K, et al. Structural similarity of bovine lung prostaglandin F synthase to lens epsilon-crystallin of the European common frog. Proc Natl Acad Sci USA 1988;85:11-15.
8. Suzuki T, Fujii Y, Miyano M, Chen L-Y, Takahashi T, Watanabe K. cDNA cloning, expression, and mutagenesis study of liver-type prostaglandin F synthase. J Biol Chem 1999;274:241-48.
9. Suzuki-Yamamoto T, Nishizawa M, Fukui M, et al. cDNA cloning, expression and characterization of human prostaglandin F synthase. FEBS Lett 1999;462(3):335-40.
10. Nishizawa M, Nakajima T, Yasuda K, et al. Close kinship of human 20alpha-hydroxysteroid dehydrogenase gene with three aldo-keto reductase genes. Genes Cells 2000;5(2):111-25.
11. Fukui M, Yasui H, Watanabe K, et al. Hypoxic contraction of contractile interstitial cells isolated from bovine lung. Am J Physiol 1996;270:L962-L72.
12. Fukui M, Fujimoto T, Watanabe K, Endo K, Kuno K. Prostaglandin F synthase is localized to contractile interstitial cells in bovine lung. J Histochem Cytochem 1996;44: 251-57.
13. Kapanci Y, Assimacopoulos A, Irle C, Zwahlen A, Gabbiani G. "Contractile interstitial cells" in pulmonary alveolar septa. J Cell Biol 1974;60:375-92.
14. Suzuki-Yamamoto T, Yokoi H, Tsuruo Y, Watanabe K, Ishimura K. Identification of prostaglandin F-producing cells in the liver. Histochem Cell Biol 1999;112(6):451-56.

DIFFERENTIAL EFFECTS OF NITRIC OXIDE ON CYCLOOXYGENASE 1 AND 2 ACTIVITIES IN CULTURED CELLS

Sei-itsu Murota, Mitsue Onodera, and Ikuo Morita

Introduction

Nitric oxide (NO) is an endogenous free radical mediator contributing to vascular homeostasis and neuronal plasticity. NO activates guanylate cyclase by binding tightly to heme Fe^{2+}, which causes the formation of nitrosyl heme. As a result, an allosteric conformational change of the apo-enzyme is induced that activates the enzyme. Because cyclooxygenase (COX) is also a heme protein, there is a possibility that NO could affect prostaglandin (PG) synthesis by a similar mechanism. However, the role of NO in PG synthesis is still considered controversial. Salvemini et al. [1] reported that NO enhanced PGE_2 production in hydronephrotic kidney. On the other hand, Tsai et al. [2] did not find any effect of NO on PG synthesis using purified cyclooxygenase. However, Kelner et al. [3] demonstrated that NO induced PGH_2/PGE_2 isomerase and enhanced PGE_2 production in NIH 3T3 cells. To explore this further, we did the following experiments in cultured cells.

Methods

CELL CULTURE

Bovine aortic endothelial cells (BAE cells) were isolated from bovine carotid artery and maintained as described previously [4]. The cells were seeded in 100 mm culture dishes with 5% FBS-MEM, and $5X10^4$ cells/ml of the cells were seeded in 24 multi-well dishes. To obtain quiescent cells, subconfluent cells were cultured with 0.3% FBS for 48 hours. To obtain activated cells, the quiescent cells were pretreated with 1 mM of aspirin for 30 minutes, followed by incubation with 10% FBS and 10ng/ml of phorbol myristate (TPA) for 6 hours.

MEASUREMENT OF COX ACTIVITY IN BAE CELLS

After 6 hours of incubation, the BAE cells were washed with PBS and incubated with 3 μg/ml of arachidonic acid in PBS for 15 minutes. COX activity was assessed by measuring

B. Samuelsson et al. (eds.), Advances in Prostaglandin and Leukotriene Research, 77–80.

6-keto$PGF_{1\alpha}$ in the reaction mixture using enzyme immunoassay kits.

COX activity in intact NIH3T3 fibroblasts was measured according to the method of Morita et al. [5]. Quiescent or serum-activated cells were incubated with 5 μM of 2', 7' -dichlorofluorescin diacetate (DCF-DA, Molecular Probe, Eugene, OR, USA) for 15 minutes. After washing with a serum-free MEM, 5 μg/ml of arachidonic acid was added, followed by measuring cellular fluorescence intensity by a fluorescence microscope with an interactive laser cytometer, adherent cell analysis and sorting (Meridian Instrument, MI, USA).

Animal Experiments

Wistar rats weighing 200-250 grams were injected intravenously with 2 mg/100 g LPS in PBS in a total volume of 100 μl per shot. Controls were injected with PBS in the same volume. Each group comprised seven rats. After 6 hours, rats were killed and aortas were isolated and washed immediately. The aorta in each rat was adjusted to the same length, and 1 ml of PBS supplemented with 3 μg/ml of arachidonic acid with or without MAHMA was perfused over the aorta for 1 minute. The perfusate was collected to measure for 6-keto $PGF_{1\alpha}$.

Results and Discussion

First, we prepared NIH 3T3 cells in which either COX-1 or COX-2 was expressed exclusively by means of the procedure described in our previous paper [6]. Briefly, when cells are cultured without the addition of serum, they express exclusively COX-1 activity. In contrast, when the cells are pretreated with aspirin to destroy preexisting COX activity and then are cultured in the presence of serum, they express only COX-2 activity.

Real-time activity of cyclooxygenase in each cell was assayed according to the new method that we established recently [5]. The cyclooxygenase activity in quiescent cells was not blocked by a COX-2 selective inhibitor NS 398, but it was abolished by aspirin, suggesting that these cells possess only COX-1 activity. On the other hand, the cyclooxygenase activity of serum-activated cells was completely blocked by NS 398 as well as aspirin, suggesting that they have only COX-2 activity. We then treated these cells with an NO donor to determine what would happen to their PG production. Interestingly, the NO donor caused a remarkable inhibition of PG synthesis in fibroblasts expressing COX-2, but not in those that express COX-1.

Next, we examined whether this interesting phenomenon demonstrated in NIH 3T3 fibroblasts could also be observed in endothelial cells. We prepared endothelial cells expressing COX-1 or COX-2 alone according to the same procedure used for NIH 3T3 fibroblasts. On the endothelial cells that express COX-1, prostacyclin production was not affected by the addition of NS-398 but was abolished by aspirin. The NO donor (NOC-7, 1 mM) had no effect on their production of prostacyclin. However, in the endothelial cells expressing only COX-2, production of prostacyclin was blocked by NS-398 as well as aspirin, as expected. In these cells, the NO donor inhibited the prostacyclin production. Thus, we obtained the same result in endothelial cells as we did in NIH 3T3 fibroblasts. The inhibitory effect of the NO donor on the COX-2 activity was dose dependent, but the NO donor did not inhibit COX-1 activity even at the higher concentrations. Thus, in COX-2 endothelial cells, prostacyclin production was blocked sharply with increasing concentration of the NO donor, while in the case of COX-1 endothelial cells, NO affected the prostacyclin production very little.

Because one of the target molecules of NO is guanylate cyclase, we examined whether guanylate cyclase was involved in the COX-2 inhibitory effect of the NO donor. Methylene blue, a specific inhibitor of guanylate cyclase, did not affect the inhibitory effect of the NO donor on COX-2 activity. In the presence of the NO donor, the production of 6-keto $PGF_{1\alpha}$ was 7% of control, whereas this value was 8% after the addition of methylene blue. In addition, the addition of 8-Br-cyclic GMP (1 mM) did not inhibit COX-2 activity.

Next we examined the effects of various kinds of NO donors, namely DETA, NOC-7, and MAHMA, each half time is 56 hours, 10.1 minutes and 2.7 minutes, respectively. When they are applied to the culture medium at the same dose, actual concentrations of NO in the culture medium were quite different from each other, i.e. the actual concentration of NO is high in the short-life NO donor, while it is low in the long-life NO donor. In all the cases, NO donors did not affect COX-1 but inhibited only COX-2 activity. Thus, regardless of their life span, NO donors inhibited only COX-2 activity. Therefore, we concluded that NO could selectively and directly inhibit the COX-2 molecule alone [7].

According to the previous paper [8], COX-2 is densely located around the nucleus, whereas COX-1 is distributed diffusely in the cytosol. There is a possibility that this difference in the intracellular localization between COX-1 and COX-2 may be responsible for the different effects of the NO donor.

To know what we obtained *in vitro* cell culture system could also apply to *in vivo*, next we examined the effect of NO on prostacyclin producing activity in aorta isolated from normal and LPS treated rats. The superfusion of aorta with exogenous arachidonic acid for 1min was carried out.

6-keto $PGF_{1\alpha}$ content in the perfusate obtained from normal rat aorta was not blocked by NS-398, suggesting that COX-1 plays a major role in prostacyclin production in normal aorta. NO donor did not affect the COX-1 activity. On the other hand, in the case of LPS-treated aorta, prostacyclin production was blocked by NS-398 suggesting that COX-2 plays a major role in this case. NO donor blocked the COX-2 activity. Thus, we obtained the same result in this *ex vivo* model as we did in the cell culture system [7].

References

1. Salvemini D, Misko TP, Masferrer JL, Seibert K, Currie MG, Needleman P. Nitric oxide activates cyclooxygenase enzymes. Proc Natl Acad Sci USA, 1993;90:7240-44.
2. Tsai AL, Wei C. Kulmacz RJ. Interaction between nitric oxide and prostaglandin H synthase. Arch Biochem Biophys 1994;313:367-72.

3. Kelner MJ, Uglik SF. Mechanism of prostaglandin E2 release and increase in PGH2/PGE2 isomerase activity by PGDF: involvement of nitric oxide. Arch Biochem Biophys 1994;312:240-43.
4. Morita I, Kanayasu T, Murota S. Kallikrein stimulates prostacyclin production in bovine vascular endothelial cells. Biochim Biophys Acta 1984;792:304-9.
5. Morita I, Smith WL, DeWitt DL, Schindler M. Expression activity profiles of cells transfected with prostaglandin endoperoxide H synthase measured by quantitative fluorescence microscopy. Biochemistry 1995;34:7194-99.
6. Shitashige M, Morita I, Murota S. Different substrate utilization between prostaglandin endoperoxide H synthase-1 and -2 in NIH3T3 fibroblasts. Biochem Biophys Acta 1998;1389:57-56.
7. Onodera M, Morita I, Mano Y, Murota S. Differential effects of nitric oxide on the activity of prostaglandin endoperoxide H synthase-1 and -2 in vascular endothelial cells. Prostaglandins, Leukot Essent Fatty Acids, 2000;62:161-67.
8. Morita I, Shindler M, Regier MK, et al. Different intracellular locations for prostaglandin endoperoxide H synthase-1 and -2. J Biol Chem 1995;270:10902-8.

On The Functional Diversity Of Secreted Phospholipases A_2: Cloning Of Novel Mammalian Enzymes And HIV-1 Antiviral Properties

Gérard Lambeau, Emmanuel Valentin, Rao Koduri, David Fenard, Alain Doglio, Michael Gelb, and Michel Lazdunski

Introduction

Over the past decade, it has become clear that mammalian cells not only express a variety of intracellular phospholipases A_2 (PLA_2), but also a diverse set of secreted phospholipases A_2 ($sPLA_2$s). While PLA_2s are generally considered as key enzymes which control the production of lipid mediators, the function of the 10 distinct $sPLA_2$s cloned so far remains ill-defined. Using venom $sPLA_2$s, two types of specific membrane receptors (N and M) have been identified in various mammalian tissues. Of physiological relevance, the M-type receptor can bind with high affinities to several mammalian $sPLA_2$s, making it likely that mammalian $sPLA_2$s are endogenous ligands of the receptors initially identified with venom $sPLA_2$s, and that the physiological function of the mammalian $sPLA_2$s is not limited to their catalytic activity.

The Mammalian $sPLA_2$ Family

To date, 9 distinct $sPLA_2$s (group IB, IIA, IIC, IID, IIE, IIF, III, V, X) have been cloned in mammals and found to have distinct, yet overlapping tissue distribution [1,2,3]. Although expressed at very high levels in pancreas, the pancreatic group IB $sPLA_2$ is also present in tissues such as lung, spleen, kidney, ovary, and prostate, suggesting functions other than dietary lipid digestion [4]. Group IIA inflammatory-type $sPLA_2$ has been detected in numerous tissues and extracellular fluids including tears, and is expressed at very high levels in various inflamed tissues and cancers [4]. This $sPLA_2$ is involved in the release of lipid mediators of inflammation and may play pivotal roles in the pathogenesis and/or progression of inflammation [4]. Group IIA $sPLA_2$ also acts as a potent antimicrobial agent released by neutrophils and able to kill various bacterial strains [4]. This enzyme may also function as a gene suppressor of tumor formation [4]. Group IIC $sPLA_2$ has been cloned from rat and mouse species but appears to correspond to a pseudogene in humans [5]. Based on its specific expression in testis, this $sPLA_2$ may have a specific function in spermatogenesis. Group V $sPLA_2$ has been cloned from humans, rat, and mouse [5]. It is mainly expressed in heart, but has recently been found in macrophages and mastocytes

B. Samuelsson et al. (eds.), Advances in Prostaglandin and Leukotriene Research, 81–84.

where it plays a role in lipid mediator release [4]. Group X $sPLA_2$ has been cloned from humans, mouse, and rat [1,6]. In humans, it is expressed in peripheral blood leukocytes, spleen, thymus, pancreas, and colon, while it is found in testis and stomach in the mouse. Group IID, IIE, IIF, and III $sPLA_2$s have recently been cloned from mouse and humans and have a specific tissue distribution pattern, suggesting distinct functions [1-3,7]. The biological function and regulation of these four later $sPLA_2$s are completely unknown.

$sPLA_2$ Receptors

Snake and insect venom also contain a large diversity of $sPLA_2$s. They share with mammalian $sPLA_2$s a similar catalytic mechanism and the same overall structural organization [8]. Several venom $sPLA_2$s are however potent toxins that exert neurotoxic, myotoxic, and proinflammatory effects [9]. In the late 1980s, it has been hypothesized that venom $sPLA_2$s exert their effects through binding to specific targets which are not recognized by nontoxic $sPLA_2$s [9]. In support to this hypothesis, we identified two types of high affinity receptors for venom $sPLA_2$s using as labeled ligand a novel venom $sPLA_2$ called OS_2 [4]. The N-type receptors are abundant in rat brain and consist of two families of binding sites with very high affinities for OS_2 (K_{d1} and K_{d2} are 1 and 45 pM, respectively). They are made of three proteins of 36-51 kDa and one protein of 85 kDa, and bind to various neurotoxic $sPLA_2$s with high affinities, suggesting the involvement of these receptors in $sPLA_2$ neurotoxic effects. Binding studies on other tissues suggest the presence of related but not identical N-type receptors [4]. The physiological role, the natural ligands, and the molecular structure of these receptors remain unknown.

The M-type (for muscle) receptor was initially identified in rabbit skeletal muscle cells, associates OS_2 with very high affinity (K_d = 7 pM), and consists of a single protein of 180 kDa [4]. This receptor is also expressed in other tissues including lung, kidney, and liver [4]. The cloned M-type receptor is a novel membrane-bound protein (1458 residues) which shares a low identity (29%) but a similar structural organization with the mannose receptor and two recently cloned receptors called DEC-205 and lectin lambda [4]. These four receptors are members of the C-type lectin superfamily and they all have a single transmembrane domain, a short cytoplasmic tail and a very large extracellular region made of a N-terminal cysteine-rich domain, a fibronectin-like type II domain and a tandem repeat of eight or ten (for DEC-205) distinct carbohydrate recognition domains called C-type CRDs. Structure-function studies indicate that the Ca^{2+}-binding loop of $sPLA_2$ binds to the CRD5 of the M-type receptor, which leads to inhibition of $sPLA_2$ activity. This finding fits well with the discovery of specific venom $sPLA_2$ inhibitors which are present in the serum of various snakes and share homology with the CRD5 of the M-type receptor [4].

The physiological role of the M-type receptor is still unclear. The M-type receptor was initially found to bind pancreatic group IB $sPLA_2$ [10], and was proposed to play a central role in various biological effects mediated by this $sPLA_2$, including cell proliferation, cell migration, and lipid mediator release [10]. Targeted disruption of the M-type $sPLA_2$ receptor gene has recently suggested that this receptor plays a critical role during the inflammatory processes induced by LPS injection and leading to endotoxic shock [11]. The endogenous ligands of the M-type receptor are still not well defined. Pancreatic group IB $sPLA_2$ was first proposed as a unique endogenous ligand of the mouse M-type receptor [11]. However, our recent study shows that both mouse group IB and mouse group IIA $sPLA_2$s are high affinity ligands for the mouse receptor, making it likely that at least these two types of $sPLA_2$s are endogenous ligands of the M-type receptor in this species [4]. The binding properties of the other $sPLA_2$s to M-type receptors remain to be analyzed.

In conclusion, it is now clear that mammalian $sPLA_2$s form a growing family of related proteins. Since most of them are catalytically active enzymes, their primary function could be linked to their enzymatic properties to release potent lipid mediators, to control bacterial infection, or to remove injured or apoptotic cells [1,2,4]. However, the use of venom $sPLA_2$s to identify $sPLA_2$ receptors that bind group IB and IIA mammalian $sPLA_2$s suggests that mammalian $sPLA_2$s can also function as high affinity ligands. Interestingly, a similar example exists for thrombin, which acts both as a proteolytic enzyme in coagulation and as a ligand through G-protein coupled receptors [12]. Finally searching for novel $sPLA_2$ functions, we recently found that venom $sPLA_2$s are potent inhibitors of HIV-1 replication through a mechanism which probably does not involve their catalytic activity, but rather involves a specific binding of $sPLA_2$ to host cells [13]. Group IB and IIA $sPLA_2$s cannot inhibit HIV-1 replication, but the inhibitory effects of the recently cloned $sPLA_2$s remain to be addressed. Clearly, understanding the physiological functions of the different members of the $sPLA_2$ family, which may be related to their catalytic activity or their binding properties, is now a challenging and important area of research.

References

1. Valentin E, Ghomashchi F, Gelb MH, Lazdunski M, Lambeau G. On the diversity of secreted phospholipases A2. Cloning, tissue distribution, and functional expression of two novel mouse group II enzymes. J Biol Chem 1999;274:31195-202.
2. Valentin E, Ghomashchi F, Gelb MH, Lazdunski M, Lambeau G. Novel human secreted phospholipase A2 with homology to the group III bee venom enzyme. J Biol Chem 2000;275:7492-96.
3. Suzuki N, Ishizaki J, Yokota Y, et al. Structures, enzymatic properties, and expression of novel human and mouse secretory phospholipase A2s. J Biol Chem 2000;275:5785-93.
4. Lambeau G, Lazdunski M. Receptors for a growing family of secreted phospholipases A2. Trends Pharmacol Sci 1999;20:162-70.
5. Tischfield J A A reassessment of the low molecular weight phospholipase A2 gene family in mammals. J Biol Chem 1997;272:17247-50.
6. Cupillard L, Koumanov K, Mattéi MG, Lazdunski M, Lambeau G. Cloning, chromosomal mapping, and expression of a novel human secretory phospholipase A2. J Biol Chem 1997;272:15745-52.
7. Valentin E, Koduri R.S, Scimeca JC, Carle G, Gelb MH, Lazdunski M, Lambeau G. Cloning and recombinant expression of a novel mouse secreted phospholipase A2. J Biol Chem 1999;274:19152-60.
8. Gelb MH, Cho W, Wilton DC. Interfacial binding of secreted phospholipases A2: more

than electrostatics and a major role for tryptophan. Curr Opin Struct Biol 1999;9:428-32.

9. Kini RM, Evans HJ. A model to explain the pharmacological effects of snake venom phospholipases A2. Toxicon 1989;27:613-35.

10. Ohara O, Ishizaki J, Arita H. Structure and function of phospholipase A2 receptor. Prog Lip Res 1995;34:117-38.

11. Hanasaki K, Yokota Y, Ishizaki J, Itoh T, Arita H. Resistance to endotoxic shock in phospholipase A2 receptor-deficient mice. J Biol Chem 1997;272:32792-97.

12. Dery O, Corvera CU, Steinhoff M, Bunnett NW. Proteinase-activated receptors: novel mechanisms of signaling by serine proteases. Am. J Physiol. 1998;274:C1429-52.

13. Fenard D, Lambeau G, Valentin E, Lefebvre JC, Lazdunski M, Doglio A. Secreted phospholipases A2, a new class of HIV inhibitors that block virus entry into host cells. J Clin Invest 1999;104:611-18.

REGULATION OF CYTOSOLIC PHOSPHOLIPASE A_2 BY PHOSPHORYLATION

Michael H. Gelb, Ying Hefner, Angelika G. Börsch-Haubold, and Steve P. Watson

Introduction

Understanding the processes by which cells liberate free arachidonic acid (AA) from phospholipids is important because this is the first step in the production of the eicosanoids, which mediate a diverse set of biological events including many inflammatory responses related to diseases. It is generally accepted that the glucocorticoid steroids exert their anti-inflammatory properties, at least in part, by blocking the cell's ability to liberate AA from membrane phospholipids. It now seems certain that the 85-kDa group IV phospholipase A_2, also known as cytosolic phospholipase A_2 (cPLA2), is directly involved in liberating AA from cellular phospholipids for the biosynthesis of eicosanoids, and thus it is important to understand how this enzyme is regulated in mammalian cells.

Regulation of cPLA2 in Mammalian Cells by Phosphorylation

Lin and co-workers were the first to provide strong evidence that cPLA2 is phosphorylated in mammalian cells in response to agonists that mobilize AA from phospholipids [1]. After this group cloned cPLA2, they realized that the protein contains a consensus motif surrounding Ser-505 for phosphorylation by mitogen-activated protein kinase family members (MAPKs). This group also provided strong evidence, but not proof, that cPLA2 is phosphorylated on Ser-505 when expressed in CHO cells. When the S505A cPLA2 mutant was expressed in CHO cells, AA release in response to Ca^{2+} ionophore and the protein kinase C activator PMA was greatly dimensioned compared to that produced by expression of wild-type cPLA2 [1]. They also showed that cPLA2 phosphorylation led to a slight decrease in the electrophoretic mobility of the enzyme. This gel shift assay, which can examined by immunoblotting methods using total protein extracted from mammalian cells, has been used by many investigators to track cPLA2 phosphorylation in response to agonists (> 300 published studies).

However, we were never completely satisfied with the gel shift assay because it does not provide site-selective cPLA2 phosphorylation data. We isolated sufficient quantities of cPLA2 from a baculovirus/insect cell expression system to determine the sites of phosphorylation by mass spectrometry, and we determined that cPLA2 was phosphorylated on Ser-505 and on Ser-437, -454, and -727 [2]. We studied cPLA2

B. Samuelsson et al. (eds.), Advances in Prostaglandin and Leukotriene Research, 85–88.

phosphorylation in platelets following stimulation with the physiological agonists thrombin and collagen. Although sufficient amounts of cPLA2 from mammalian cells for mass spectrometry studies cannot be readily obtained, we were able to use ^{32}P-methods to show that the radiolabeled platelet cPLA2 tryptic peptides chromatographically co-migrated with authentic phosphopeptides. We showed that cPLA2 is stoichiometrically phosphorylated in thrombin- and collagen-stimulated platelets on Ser-505 and on Ser-727, and this accounted for all of the phosphorylation in these [3]. We also showed that cPLA2 is phosphorylated exclusively on Ser-505 and on Ser-727 in HeLa cells stimulated with activators of the stress-activated protein kinases (SAPKs, members of the MAPK family) [3]. Our work cast doubt on the suggestion, based only on immunoblot analysis, that cPLA2 phosorphylated on tyrosine in HeLa cells {Flati, #2578}[4].

Kramer and co-workers showed that the p38 member of the SAPK family or a closely related kinase and not the p42/p44 MAPKs (ERK1 and 2) was responsible for cPLA2 phosphorylation in platelets [5]. We found that the inhibitors of p38, SB202190 and SB203580, blocked about 50% of the phosphorylation on both Ser-505 and Ser-727 in thrombin-stimulated platelets. The inhibition data suggests that the kinase responsible for Ser-727 phosphorylation is activated by p38. The fact that cPLA2 phosphorylation is only 50% inhibited by SAPK inhibitors suggests that more than one isoform of p38 (inhibitor sensitive and insensitive forms) may be phosphorylating cPLA2 in platelets, and this was shown to be the case [6]. Interestingly, in collagen-stimulated platelets, only the SB202190/SB203580-sensitive SAPKs, SAPK2a/2b, become activated, and we found that these inhibitors block all of the cPLA2 phosphorylation (Ser-505 and Ser-727) induced by collagen [6]. We showed that complete blockage of cPLA2 phosphorylation leads to a shift in the AA release versus collagen concentration dose-response curve in platelets. Three-fold higher amounts of collagen are needed to elicit the same AA release response in the absence of cPLA2 phosphorylation (i.e. with SB202190/SB203580 present) as in the presence of phosphorylation [6].

Our studies with SAPK inhibitors SB202190/SB203580 led us to propose that the kinase responsible for phosphorylation of cPLA2 on Ser-727 is activated by one or more SAPKs. We decided to test all of the kinases known at the time of our study to be activated by SAPKs. Using mass spectrometry, we proved that MNK1 phosphorylates cPLA2 *in vitro* uniquely at Ser-727 and that PRAK1 and MSK1 produce a radiophosphorylated tryptic peptide that co-migrates with authentic Ser-727 phosphorylated tryptic peptide chromatographically [7]. MAPKAP-K2 and MAPKAP-K3 did not phosphorylate cPLA2 *in vitro*.

We further examined the functional consequence of Ser-505 and Ser-727 cPLA2 phosphorylation [7]. Stable CHO cell transfectants were prepared that express wild type and mutant forms of cPLA2. As shown in Figure 1, expression of wild type cPLA2 in CHO cells leads to an increase in AA release compared to the nontransfected parental cells. Stimulation with Ca^{2+} ionophore alone leads to more arachidoante release than stimulation with the protein kinase C agonist PMA alone, and there is a synergistic effect when both agonists are added. Mutation of either Ser-505 or Ser-727 significantly reduces AA release, and the double mutant fails to produce AA above the level made by nontransfected cells. Interestingly, when a high concentration of Ca^{2+} ionophore is used, the effect of double mutation on arachidonate release is rescued (Figure 1, right panel) showing that the mutant can be fully functional in these cells. These results establish that phosphorylation of cPLA2 on Ser-505 and on Ser-727 is important for efficient AA release and that high Ca^{2+} concentrations can overcome the need for cPLA2 phosphorylation. The results also underscore the need for rigorous analysis of cPLA2 phosphorylation using site-selective protein chemical techniques rather than simply relying on gel shift analysis.

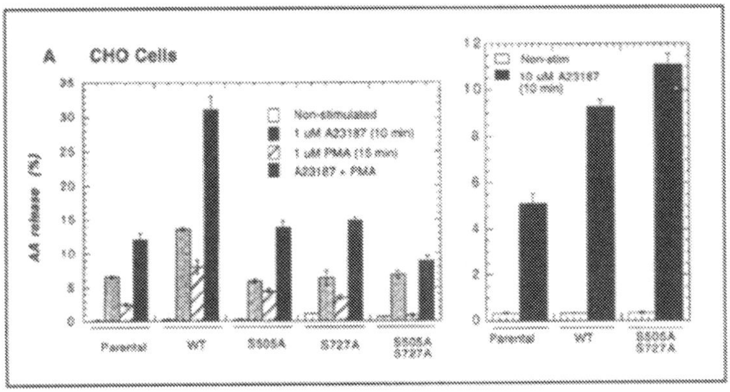

Figure 1. AA release from A23187 and PMA stimulated nontransfected CHO cells (parental) or cells transfected with wild type (WT) and cPLA2 mutants (S505A, S727A, and double mutant). Right panel is with high A23187 and only 1 minute of stimulation.

References

1. Lin L-L, Wartmann M, Lin AY, Knopf JL, Seth A, Davis RJ. cPLA2 is phosphorylated and activated by MAP kinase. Cell 1993;72:269-78.
2. de Carvalho MGS, McCormack AL, Olson E, et al. Identification of phosphorylation sites of human 85-kDa cytosolic phospholipase A_2 expressed in insect cells and present in human monocytes. J Biol Chem 1996;96:6987-97.
3. Börsch-Haubold AG, Bartoli F, Asselin J, et al. Identification of the phosphorylation sites of cytosolic phospholipase A_2 in agonist-stimulated human platelets and HeLa cells. J Biol Chem 1998;273:4449-58.
4. Flati V, Haque SJ, Williams BR. Interferon-alpha-induced phosphorylation and activation of cytosolic phospholipase A_2 is required for the formation of interferon-stimulated gene factor three. EMBO J 1996;15:1566-71.
5. Börsch-Haubold AG, Kramer SP, Watson SP. Phosphorylation and activation of cytosolic phospholipase A_2 by 38-kDa mitogen-activated protein kinase in collagen-stimulated human platelets. J Biol Chem 1997;245:751-59.
6. Börsch-Haubold AG, Ghomashchi F, Pasquet S, et al. Phosphorylation of cytosolic

phospholipase A_2 in platelets is mediated by multiple stress-activated protein kinase pathways. Eur J Biochem 1999;265:195-203.

7. Hefner Y, Börsch-Haubold AG, Wilde JI, et al. Serine-727 phosphorylation and activation of cytosolic phospholipase A_2 by MNK1-related protein kinases. J Biol Chem 2000, in press.

Synthesis and Function of 20-Hydroxyeicosatetraenoic Acid in the Kidney

Michal Laniado Schwartzman, Jackleen Marji, Miao Jiang, and Mong-Heng Wang

Introduction

20-hydroxyeicosatetraenoic acid (20-HETE), the ω-hydroxylation product of arachidonic acid (AA), is the principal metabolite formed in tubular and vascular structures of the rat renal cortex and outer medulla. 20-HETE has potent biological activities and has been shown to contribute to the regulation of renal function and to the control of arterial pressure. In the renal tubules it inhibits sodium reabsorption, while in the renal microcirculation it is a vasoconstrictor and a regulator of the myogenic response. The ω-hydroxylation of fatty acids, including AA, is catalyzed by enzymes of the cytochrome P450 (CYP) 4A family. In the rat, four isoforms have been identified: CYP4A1, 4A2, 4A3, and 4A8. Our studies indicated that despite the high homology, these isoforms display distinct catalytic properties including differences in kinetic parameters, product profile and inhibitor sensitivity. While the constitutive level of expression of CYP4A1 is low, its recombinant form is the low K_m AA-ω-hydroxylase and thus, by far, the most efficient 20-HETE synthesizing enzyme. Whereas CYP4A1 is solely an AA ω-hydroxylase, CYP4A2 and CYP4A3 also catalyze AA 11,12-epoxidation. Systemic administration of CYP4A antisense oligodeoxynucleotides revealed that CYP4A1 contributes significantly to the renal tubular and vascular production of 20-HETE. Furthermore, these isoforms demonstrate unique intrarenal localization. Using molecular and pharmacological probes we demonstrated a unique CYP4A isoform-specific localization within the renal microvasculature. Transfection of CYP4A1 cDNA to renal microvessels resulted in enhanced 20-HETE synthesis and increased reactivity to phenylephrine. Thus, 20-HETE of vascular origin serves as a stimulatory regulator of vascular responses to constrictor agonists.

Renal CYP4A Expression and Activity

The ω-hydroxylation of fatty acids, including AA, is catalyzed primarily by enzymes of the CYP4A gene family. This subfamily encodes several CYP enzymes that are capable of hydroxylating the terminal ω-carbon and, to a lesser extent, the (ω-1) position of saturated and unsaturated fatty acids as well as various prostaglandins [1]. In the rat, four members of this family have been identified: CYP4A1, CYP4A2, CYP4A3, and CYP4A8

B. Samuelsson et al. (eds.), Advances in Prostaglandin and Leukotriene Research, 89–97.

and messages for all four have been identified in the kidney [2-4]. These isoforms, although sharing 66-98% homology and a common, unique catalytic activity, i.e. hydroxylation of fatty acid at the ω-carbon, are localized to different renal structures and are exposed to different regulatory mechanisms. For example, whereas CYP4A2 is preferentially expressed in the outer-medullary/mTAL region, CYP4A1, CYP4A3, and CYP4A8 are highly expressed in the proximal tubules [4-6]. In addition, CYP4A2 is believed to be the major CYP4A isoform expressed in the renal microvasculature, a major site of 20-HETE synthesis and action [7]. Renal CYP4A1 and CYP4A3 can be induced by hypolipidemic drugs such as clofibrate, whereas CYP4A2 is thought to be constitutively expressed especially in male rats [2,3,8,9]. The CYP4A8 isoform, originally isolated from rat prostate, exhibits 76% and 71% amino acid homology with CYP4A1 and CYP4A2/CYP4A3, respectively. It is distributed along the outer stripe of the outer medulla, indicating its localization in the proximal tubule [10]. A recent study by Ito et al. [11] examined the expressions of CYP4A isoforms in various nephron segments and preglomerular arterioles microdissected from the kidneys of Sprague-Dawley rats. Whereas CYP4A2, CYP4A3, and CYP4A8 mRNAs are constitutively expressed, CYP4A1 mRNA is barely detectable. The expression of CYP4A1, CYP4A3, and CYP4A8 is similar in both male and female kidneys; however, CYP4A2 expression was four-fold greater in the kidney of male versus female rats. Thus, the expression of CYP4A isoforms is sex dependent and different isoforms are expressed throughout various nephron segments and the renal vasculature of the rats.

We found that despite the high homology these isoforms display different catalytic properties (Table 1). Thus, recombinant CYP4A1, CYP4A2, and CYP4A3, but not CYP4A8, catalyzed AA ω-hydroxylation to 20-HETE with the highest catalytic efficiency (V_{max}/K_m) for CYP4A1, followed by CYP4A2 and CYP4A3. Moreover, CYP4A2 and CYP4A3 exhibited an additional AA-11,12-epoxidation activity (formation of 11,12-EET) whereas CYP4A1 operated solely as a ω-hydroxylase [12,13].

Table 1. Initial velocity kinetic constants of baculovirus expressed CYP4A isoforms with AA. The V_{max} and K_m values were determined by linear regression of Lineweaver-Burke plots. Results are the mean of three separate determinations, SE < 10%.

	V_{max} (min^{-1})	K_m (μM)	V_{max}/K_m (nM^{-1} min^{-1})
CYP4A1	9.47	10	947
CYP4A2	1.38	19	72
CYP4A3	0.92	41	22
CYP4A8	ND	ND	ND

The renal expression of these isoforms is also age-dependent. CYP4A1 and CYP4A3 proteins are detectable in the fetus and their levels are gradually increased from newborn until about 9 weeks of age and then the levels decline to very low levels in adults. CYP4A2 protein is undetectable until 5 weeks but then the levels increase such that in adult male rat it is the major isoform detected in the kidney [14,15]. CYP4A8 mRNA levels are detected at 3 weeks of age and thereafter follow a similar pattern of expression to that of CYP4A1 and CYP4A3 [15]. Renal cortical CYP4A1 protein levels were higher in SHR compared to SD and WKY rats. The increased levels of CYP4A1 proteins at 7 weeks corresponded to the maximal activity of AA ω-hydroxylation suggesting that this isoform may play a significant role in contributing to the increased cortical production of 20-HETE seen in 7-week-old SHR rats [14].

The involvement of these proteins/genes in the pathogenesis of hypertension is not only implied from the biological effects of their major catalytic product, 20-HETE, but also from studies of their expression in animal models of hypertension. To this end, the CYP4A2 gene is one of the few genes preferentially expressed in the kidney of young SHR [6,16] where renal 20-HETE synthesis is high [17]. Depletion of renal CYP or inhibition of renal CYP activities decreases 20-HETE synthesis and lowers blood pressure in the SHR [18-20]. Moreover, CYP4A genotype has been reported to cosegregate with the development of salt-induced hypertension in F_2 populations derived from a cross of both SHR [21] and Dahl S rats [22] with normotensive strains. Recent studies in our lab demonstrated that treatment of rats with CYP4A antisense oligodeoxynucleotides inhibited renal vascular synthesis and urinary excretion of 20-HETE and significantly reduced blood pressure in SHR and SD rats (Figure 1) [23]. On the other hand, clofibrate treatment which readily induces CYP4A1 and CYP4A3 (but not CYP4A2) expression, increased outermedullary 20-HETE and lowers blood pressure in female Dahl SS rats [24]. Collectively, these studies provide strong evidence that the CYP4A genes and the catalytic activity of their products (primarily 20-HETE) are important factors in the development and/or maintenance of hypertension.

Renal 20-HETE Synthesis and Effects

20-HETE was found to constitute a major CYP AA metabolite in tubular and vascular structures of the renal cortex and outer medulla of the rat [7,25]. 20-HETE formation is localized in the proximal tubules (segments S1, S2, S3) [25,26], medullary thick ascending limb [27], preglomerular microvessels [7], and glomeruli [28]. We have showed that more than 85% of renal 20-HETE is formed in the proximal tubules with the highest production in the S3 segment [25]. 20-HETE has been detected in the human and rat urine [29,30] indicating that it is an endogenous constituent. The ω-hydroxylation of AA, i.e. formation of 20-HETE, was though to be a catabolic step in the oxidation and degradation of this fatty acid. Our studies were the first to characterize 20-HETE as a potent vasoconstrictor eicosanoid [31,32].

20-HETE exhibits potent renal effects and its localization within the kidney enables easy access to the systemic circulation as well as to the intrarenal arteries suggesting a possible role of 20-HETE in the regulation of renal function in autocrine and/or paracrine fashion. The major vascular (vasoconstriction [7,33-35]) and tubular (inhibition of ion transport [36]) effects of 20-HETE are believed to be the mechanisms by which it affects autoregulation of renal blood flow, tubuloglomerular feedback, and salt and water reabsorption [37-41]. Its role in the regulation of blood pressure has been further implied from studies showing that inhibition and stimulation of its formation affect arterial pressure [18,24,42].

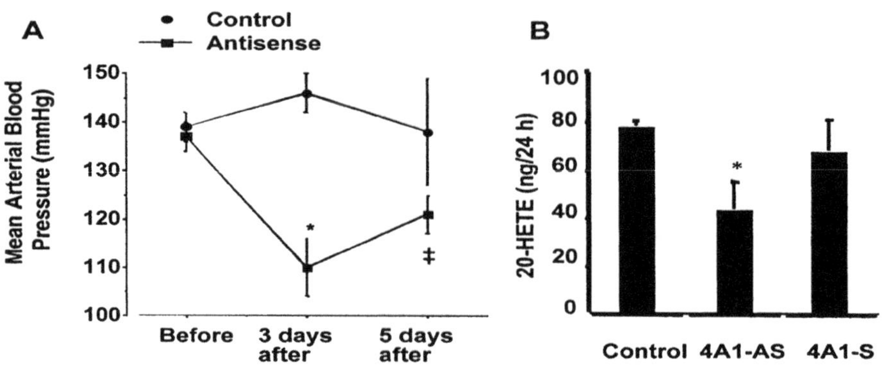

Figure 1. Effect of CYP4A1 antisense (AS) and sense (S) oligodeoxynucleotides on blood pressure (A) and urinary 20-HETE excretion (B). ODNs (4A1-AS or 4A1-S, each at a dose of 167 nmol.kg body wt $^{-1}$.day $^{-1}$ for 5 days) or vehicle control were administered into the femoral vein of 7-week-old SHR. Blood pressure was measured before and after 3 and 5 days of treatment by the tail cuff method. Results are mean ± SE, n=4; * $p < 0.05$ and ‡ $p < 0.001$ from control. Twenty-four hour urinary 20-HETE was quantitated by NCI-GC/MS analysis; n=4. $p < 0.05$ from control.

The cellular mechanisms underlying the potent biological activities of 20-HETE are still unclear. 20-HETE increases intracellular Ca^{2+} concentration in renal arteriolar vascular smooth muscle cells presumably via inhibition of the large-conductance Ca^{2+}-activated K^+ channel activity [43]. Elevation of intracellular Ca^{2+} concentrations may constitute a common second messenger system by which 20-HETE induces vascular smooth muscle constriction and stimulates renal epithelial cell growth [44]. In mTAL, 20-HETE inhibits K^+ efflux via a large conductance (70 ps) K^+ channel [45]. This limits the amount of K^+ available for transport via the Na^+-K^+-$2Cl^-$ cotransporter and reduces the lumen positive potential, which is the main driving force for passive cations reabsorption in this portion of the nephron. Previous studies have also demonstrated that 20-HETE can activate protein kinase C (PKC), which is involved in the regulation of Na^+/K^+-ATPase activity [46]. The 20-HETE-dependent inhibition of rat renal Na^+/K^+-ATPase was abolished not only by specific PKC inhibitors, but also by mutation of the $\alpha 1$ regulatory site. This evidence suggested that 20-HETE may act as a second messenger, mediating the actions of hormones on phosphorylation of rat renal α1 Na^+/K^+-ATPase by PKC. In addition, 20-HETE-induced vasoconstriction and inhibition of K^+ current in cat cerebral vascular smooth muscle have been found to be dependent on the activation of PKC [47]. Recently, Muthalif et al. has demonstrated that 20-HETE can increase mitogen-activated protein kinase (MAPK) and cytosolic PLA_2 activities, thus stimulating AA release and causing translocation of Ras in rabbit vascular smooth muscle cells [48]. They proposed that the signaling mechanism by which norepinephrine, AII, and EGF activate the Ras/MAPK pathway might be through the generation of 20-HETE. Tyrosine kinase and PKC have both been implicated in the vasoconstrictor response to 20- HETE in renal arterioles [49]. A recent study by Alonso-Galicia et al. [50] suggested the existence of a binding site or receptor for 20-HETE in renal microvessels. Whether this receptor/binding site is a channel or kinase or other structural protein is remained to be explored.

Several hormones were found to modulate kidney function by controlling the synthesis and release of 20-HETE. In the rabbit, angiotensin (AII) was found to increase renal efflux of 20-HETE [51]. Endothelin (ET-1) evoked an increase in 20-HETE release from the rat kidney and inhibition of CYP-dependent AA metabolism greatly reduced the renal vascular response to ET-1 [52]. These studies suggested that 20-HETE may mediate the renal vasoconstrictor responses to both ET-1 and AII. In addition, parathyroid hormone (PTH) and EGF were found to stimulate the production of 20-HETE [25], which is thought to serve as a mediator of PTH-induced natriuresis and EGF-stimulated mitogenic activity [44,53]. More recently, several reports have shown that the inhibitory action of angiotensin II and bradykinin on sodium transport in the mTAL can be blocked by inhibitors of CYP suggesting that they may work by stimulating intracellular production of 20-HETE [54,55]. For example, the inhibitory effect of low concentration of AII (5×10^{-11} M) on the apical K^+ channel was found to be mediated via the 20-HETE pathway [27]. In addition, very low AII concentrations ($\leq 10^{-12}$ M) through high-affinity receptor occupancy maximally inhibits the Na^+-K^+-$2Cl^-$ cotransporter via diacylglycerol-lipase and CYP-derived products (i.e. 20-HETE) [54]. The addition of 20-HETE to the bathing medium of the rat microperfused mTAL elicited a significant reduction in Cl^- transport suggesting that the bradykinin-dependent inhibition of NaCl transport in this region of the nephron can be mediated by CYP-dependent AA metabolites [55]. Altogether, these studies suggested that 20-HETE plays an important role in many key elements of modulating renal function (Figure 2).

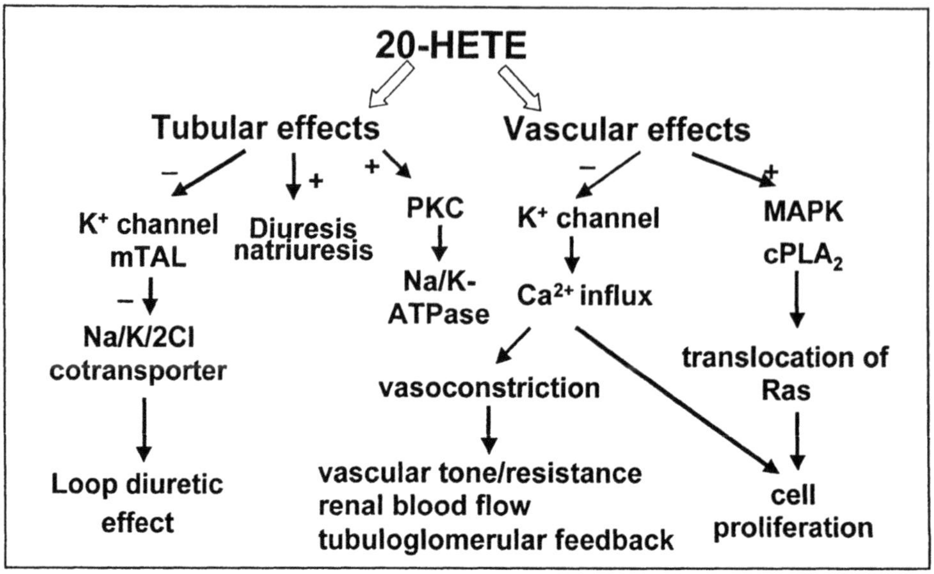

Figure 2. Biological activities of 20-HETE in the kidney and their consequences on renal function. Stimulation is denoted as +; inhibition is denoted as -.

Acknowledgment

This study was supported, in part, by a grant from the National Heart, Lung and Blood Institute, PO1 34300 and by the American Heart Association Grant to Mong-Heng Wang (99-30277T).

References

1. Gibson GG. Comparative aspects of the mammalian cytochrome P450 IV gene family. Xenobiotica 1989;19:1123-48.
2. Kimura S, Hanioka N, Matsunaga E, Gonzalez FJ. The rat clofibrate-inducible CYP4A gene subfamily I. Complete intron and exon sequence of the CYP4A1 and CYP4A2 genes, unique exon organization, and identification of a conserved 19-bp upstream element. DNA 1989;8:503-16.
3. Kimura S, Hardwick JP, Kozak CA, Gonzalez FJ. The rat clofibrate-inducible CYP4A gene subfamily. II. cDNA sequence of IVA3, mapping of the Cyp4a locus to mouse chromosome 4, and coordinate and tissue-specific regulation of the CYP4A genes. DNA 1989;8:517-25.
4. Stromstedt M, Hayashi SI, Zaphiropoulos PG, Gustafsson JA. Cloning and

characterization of a novel member of the cytochrome P450 subfamily IVA in rat prostate. DNA 1990;9:567-77.

5. Hardwick JP. CYP 4A subfamily: functional analysis by immunocytochemistry and in situ hybridization. Meth Enzymol 1991;206:273-83.
6. Iwai N, Inagami T. Isolation of preferentially expressed genes in the kidneys of hypertensive rats. Hypertension 1991;17:161-69.
7. Imig JD, Zou A-P, Stec DE, Harder DR, Falck JR, Roman RJ. Formation and actions of 20-hydroxyeicosatetraenoic acid in rat renal arterioles. Am J Physiol 1996;270:R217-R227.
8. Imaoka S, Yamazoe Y, Kato R, Funae Y. Hormonal regulation of rat cytochrome P450s by androgen and pituitary. Arch Biochem Biophys 1992;299:179-84.
9. Sundseth SS, Waxman DJ. Sex-dependent expression and clofibrate inducibility of cytochrome P450 4A fatty acid w-hydroxylases. J Biol Chem 1993;267:3915-21.
10. Stromstedt M, Warner M, Gustafsson JA. Cytochrome P450s of the 4A subfamily in the brain. J Neurochem 1994;63:671-76.
11. Ito O, Alonso-galicia M, Hopp KA, Roman RJ. Localization of cytochrome P-450 4A isoforms along the rat nephron. Am J Physiol 1998;274:F395-F404.
12. Wang M-H, Stec DE, Balazy M, et al. Cloning, sequencing and cDNA-directed expression of the rat renal CYP4A2: arachidonic acid w-hydroxylation and 11,12-epoxidation by CYP4A2 protein. Arch Biochem Biophys 1996;336:240-50.
13. Nguyen X, Wang MH, Reddy KM, Falck JR, Schwartzman ML. Kinetic profile of the rat CYP4A isoforms: arachidonic acid metabolism and isoform-specific inhibitors. Am J Physiol 1999;276:R1691-R1700.
14. Laniado Schwartzman M, Da Silva J-L, Lin F, Nishimura M, Abraham NG. Cytochrome P450 4A expression and arachidonic acid w-hydroxylation in the kidney of the spontaneously hypertensive rat. Nephron 1996;73:652-63.
15. Kroetz DL, Huse LM, Thuresson A, Grillo MP. Developmentally regulated expression of the CYP4A genes in the spontaneously hypertensive rat kidney. Mol Pharmacol 1997; 52:362-72.
16. Imaoka S, Funae Y. Hepatic and renal cytochrome P450s in spontaneously hypertensive rats. Biochim Biophys Acta 1991;1074:209-13.
17. Omata K, Abraham NG, Escalante B, Laniado Schwartzman M. Age-related changes in renal cytochrome P450 arachidonic acid metabolism in spontaneously hypertensive rats. Am J Physiol 1992;262:F8-F16.
18. Sacerdoti D, Escalante B, Abraham NG, McGiff JC, Levere RD, Schwartzman ML. Treatment with tin prevents the development of hypertension in spontaneously hypertensive rats. Science 1989;243:388-90.
19. Levere RD, Martesak P, Escalante B, Schwartzman ML, Abraham NG. Effect of heme arginate administration on blood pressure in spontaneously hypertensive rats. J Clin Invest 1990;86:213-19.
20. Su P, Kaushal KM, Kroetz DL. Inhibition of renal arachidonic acid omega-hydroxylase activity with ABT reduces blood pressure in the SHR. Am J Physiol 1998;275:R426-R438.
21. Stec DE, Trolleit MR, Krieger JE, Jacob HJ, Roman RJ. Cytochrome P4504A activity and salt sensitivity in spontaneously hypertensive rats. Hypertension 1996;27:1329-36.
22. Stec DE, Deng AY, Rapp JP, Roman RJ. Cytochrome P4504A genotype cosegregates

with hypertension in Dahl S rats. Hypertension 1996;27:564-68.

23. Wang M-H, Guan H, Nguyen X, Zand B, Nasjletti A, Laniado-Schwartzman M. Contribution of cytochrome P450 4A1 and 4A2 to vascular 20-hydroxyeicosatetraenoic acid synthesis in the rat kidney. Am J Physiol 1998;276:F246-F253.

24. Roman RJ, Ma Y-H, Frohlich B, Markham B. Clofibrate prevents the development of hypertension in Dahl salt-sensitive rats. Hypertension 1993;21:985-88.

25. Omata K, Abraham NG, Laniado Schwartzman M. Arachidonic acid w/w-1 hydroxylase along the nephron of the spontaneously hypertensive rat. Am J Physiol 1992;262:F592-99.

26. Laniado Schwartzman M, Abraham NG. The renal cytochrome P450 arachidonic acid system. Pediat Nephrol 1992;6:490-98.

27. Lu M, Zhu Y, Balazy M, Reddy KM, Falck JR, Wang W-H. Effect of angiotensin II on the apical K+ channel in the thick ascending limb of the rat kidney. J Gen Physiol 1996; 108:537-47.

28. Ito O, Roman RJ. Regulation of P-450 4A activity in the glomerulus of the rat. Am J Physiol 1999;76:R1749-R1757.

29. Laniado Schwartzman M, Omata K, Lin F, Bhatt RK, Abraham NG. Detection of 20-hydroxyeicosatetraenoic acid in rat urine. Biochem Biophys Res Commun 1991;180:445-49.

30. Prakash C, Zang JY, Falck JR, Chauhan K, Blair IA. 20-hydroxyeicosatetraenoic acid is excreted as a glucuronide conjugate in human urine. Biochem Biophys Res Commun 1992;185:728-33.

31. Escalante B, Sessa WC, Falck JR, Yadagiri P, Schwartzman ML. Vasoactivity of 20-hydroxyeicosatetraenoic acid is dependent on metabolism by cyclooxygenase. J Pharmacol Exp Ther 1988;248:229-32.

32. Laniado Schwartzman M, Falck JR, Yadagiri P, Escalante B. Metabolism of 20-hydroxyeicosatetraenoic acid by cyclooxygenase: formation and identification of novel endothelium-dependent vasoconstrictor metabolites. J Biol Chem 1989;264:1165-62.

33. Escalante B, Omata K, Sessa W, lee S-G, Falck JR, Laniado-Schwartzman M. 20-hydroxyeicosatetraenoic acid is an endothelium-dependent vasoconstrictor in rabbit arteries. Eur J Pharmacol 1993;235:1-7.

34. Ma Y-H, Gebremedhin D, Schwartzman ML, et al. 20-hydroxyeicosatetraenoic acid is an endogenous vasoconstrictor of canine renal arcuate arteries. Circ Res 1993;72:126-36.

35. Imig JD, Falck JR, Roman RJ. 20-hydroxyeicosatetraenoic acid (20-HETE) is a potent endogenous vasoconstrictor of renal arterioles. FASEB J 1993;7:A888.

36. Escalante B, Erlij D, Falck JR, McGiff JC. Cytochrome P450 arachidonate metabolites affect ion fluxes in rabbit medullary thick ascending limb. Am J Physiol 1994;266: C1775-C1782.

37. Imig JD, Zou A-P, Ortiz-de-Montellano PR, Sui Z, Roman RJ. Cytochrome P450 inhibitors alter afferent arteriolar responses to elevations in pressure. Am J Physiol 1994; 266:H1879-H1885.

38. Zou A-P, Imig JD, Kaldunski M, Ortiz de Montellano PR, Sui Z, Roman RJ. Inhibition of renal vascular 20-HETE production impairs autoregulation of renal blood flow. Am J Physiol 1994;266:F275-F282.

39. Zou A-P, Imig JD, Ortiz de Montellano PR, Sui Z, Falck JR, Roman RJ. Effect of P450 w-hydroxylase metabolites of arachidonic acid on tubuloglomerular feedback. Am J

Physiol 1994;266:F934-F941.
40. Iterman MA, Chaurasia CS, Lu P, Hardwick JP, Hanzlik RP. Fatty acid discrimination and w-hydroxylation by cytochrome P450 4A1 and cytochrome P4504A1/NADPH-P450 reductase fusion protein. Arch Biochem Biophys 1995;320:289-96.
41. Zou A-P, Drummond HA, Roman RJ. Role of 20-HETE in elevating loop chloride reabsorption in Dahl SS/Jr rats. Hypertension 1996;27:631-5.
42. Gebremedhin D, Ma Y-H, Imig JD, Harder DR, Roman RJ. Role of cytochrome P450 in elevating renal vascular tone in spontaneously hypertensive rats. J Vasc Res 1993;30:53-60.
43. Zou A-P, Fleming JT, Falck JR, et al. 20-HETE is an endogenous inhibitor of the large-conductance Ca^{2+}-activated K^+ channel in renal arterioles. Am J Physiol 1996;270:R228-R237.
44. Lin F, Rios A, Falck JR, Belosludtsev Y, Laniado Schwartzman M. 20-hydroxyeicosa-tetraenoic acid is formed in response to EGF and is a mitogen in rat proximal tubule. Am J Physiol 1995; 269:F806-F816.
45. Wang W, Lu M. Effect of arachidonic acid on activity of the apical K^+ channel in the thick ascending limb of the rat kidney. J Gen Physiol 1995;106:727-43.
46. Nowicki S, Chen SL, Aizman O, et al. 20-Hydroxyeicosa-tetraenoic acid (20 HETE) activates protein kinase C. Role in regulation of rat renal Na^+,K^+-ATPase. J Clin Invest 1997;99:1224-30.
47. Lange A, Gebremedhin D, Narayanan J, Harder D. 20-Hydroxyeicosatetraenoic acid-induced vasoconstriction and inhibition of potassium current in cerebral vascular smooth muscle is dependent on activation of protein kinase C. J Biol Chem 1997;272:27345-52.
48. Muthalif MM, Benter IF, Karzoun N, et al. 20-Hydroxyeicosatetraenoic acid mediates calcium/calmodulin-dependent protein kinase II-induced mitogen-activated protein kinase activation in vascular smooth muscle cells. Proc Natl Acad Sci USA 1998;95: 12701-6.
49. Sun CW, Falck JR, Harder DR, Roman RJ. Role of tyrosine kinase and PKC in the vasoconstrictor response to 20- HETE in renal arterioles. Hypertension 1999;33:414-18.
50. Alonso-Galicia M, Falck JR, Reddy KM, Roman RJ. 20-HETE agonists and antagonists in the renal circulation. Am J Physiol 1999;277:F790-F796.
51. Carroll MA, Balazy M, Huang DD, Rybalova S, Falck JR, McGiff JC. Cytochrome P450-derived renal HETEs: storage and release. Kidney Int 1997;51:1696-702.
52. Oyekan AO, McGiff JC. Cytochrome P-450-derived eicosanoids participate in the renal functional effects of ET-1 in the anesthetized rat. Am J Physiol 1998;274:R52-R61.
53. Pedrosa CM, Dubay GR, Falck JR, Mandel LJ. Parathyroid hormone inhibits Na^+-K^+-ATPase through a cytochrome P450 pathway. Am J Physiol 1994;266:F497-F505.
54. Amlal H, LeGoff C, Vernimmen C, Soleimani M, Paillard M, Bichara M. ANG II controls Na($^+$)-K^+(NH^{4+})-2Cl- cotransport via 20-HETE and PKC in medullary thick ascending limb. Am J Physiol 1998;274:C1047-C1056.
55. Grider JS, Falcone JC, Kilpatrick EL, Ott CE, Jackson BA. P450 arachidonate metabolites mediate bradykinin-dependent inhibition of NaCl transport in the rat thick ascending limb. Can J Physiol Pharmacol 1997;75:91-96.

ALTERED BRONCHODILATION AND PULMONARY INFLAMMATION IN PROSTANOID EP_2 RECEPTOR KNOCKOUT MICE

Francois Nantel, Danielle Denis, Simon Wong, Paul Masson, Denis Normandin, Maria Cirino, Gary P. O'Neill, Richard M. Breyer, and Kathleen M. Metters

Introduction

Small clinical studies of asthma [1-5] found that inhaled prostaglandin E_2 (PGE_2) induced bronchodilation and reduced the influx of inflammatory cells in the airways. PGE_2 activates four receptors identified as EP_{1-4} [6,7]. The activation of two of those receptors, the EP_2 and the EP_4, leads to the stimulation of the adenylyl cyclase and increases intracellular cyclic AMP levels. Therefore, a specific EP_2 or EP_4 agonist could be considered as a potential treatment for asthma.

We evaluated the pulmonary phenotype of mice in which the prostanoid EP_2 receptor was inactivated by homologous recombination [8].

Results and Discussion

The potential role of the EP_2 receptor in bronchodilation was addressed by investigating the ability of inhaled PGE_2 to inhibit carbachol-induced bronchoconstriction. Carbachol induced a significant increase of PenH in WT mice (Figure 1). When those mice were pre-treated with aerosolized PGE_2 (300 μg/ml) for 1 hour before carbachol, this caused a 10-fold rightward shift in the carbachol dose-response curve and a reduction in the maximal response. In KO mice, the maximal response to carbachol was approximately 50% greater than in WT mice. Moreover, pretreatment of these KO mice with PGE_2 had no significant effect on the carbachol dose-response curve. We can therefore conclude from this experiment that the bronchodilatory actions of PGE_2 are indeed mediated through activation of the EP_2 receptor.

To investigate the role of the EP_2 receptor in allergen-induced inflammation, WT and KO mice were sensitized with two i.p. injection of ovalbumin (OVA, 8μg) over a 12-day period and subsequently challenged for 1 hour with aerosolized OVA (1 mg/ml). Histological evaluation of the lungs showed a significant influx of inflammatory cells in the WT mice at 24 and 72 hours post-challenge. However, we observed reduction of the extent of inflammation in KO lungs (Figure 2a,b). Moreover, the pulmonary inflammation of WT mice was accompanied by elevated levels of IL-4 and IL-5 (Figure 2c,d) in the BAL

B. Samuelsson et al. (eds.), Advances in Prostaglandin and Leukotriene Research, 99–101.

fluid and IL-5 in serum (Figure 2e) that were not observed in the KO mice. These results indicate that the absence of EP_2 receptors in the KO mice leads a reduced inflammatory response to antigen challenge. Both WT and KO serum showed similar levels of OVA-specific IgE (15.4 ± 4.9 and 16.5 ± 6.1 ng/ml, respectively, n=8) following the sensitization procedure with OVA. Therefore, the lower level of inflammation of the KO did not result from differences in the sensitization to the allergen.

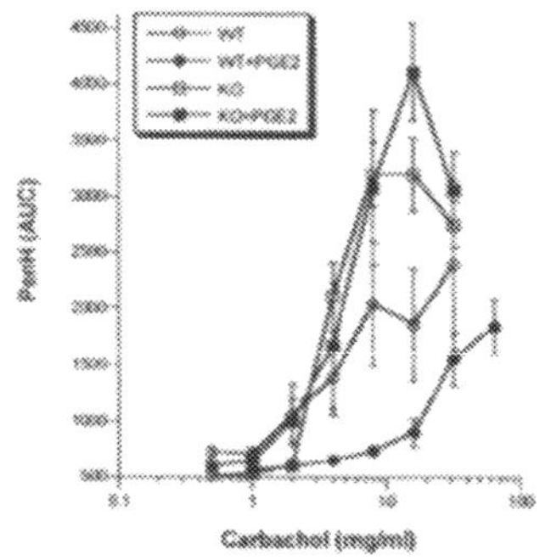

Figure 1. Effect of PGE_2 on carbachol-induced bronchoconstriction. WT and KO were placed in a whole body plethysmograph and exposed for 3 minutes to increasing doses of carbachol (0.5 to 64 mg/ml). A minimum of 15 minutes, or until PenH value return within 10% of baseline, was allowed between doses. Five minutes before each dose animals received a 1 minute aerosol of either saline or PGE_2 (300 μg/ml). n=4-8 animals.

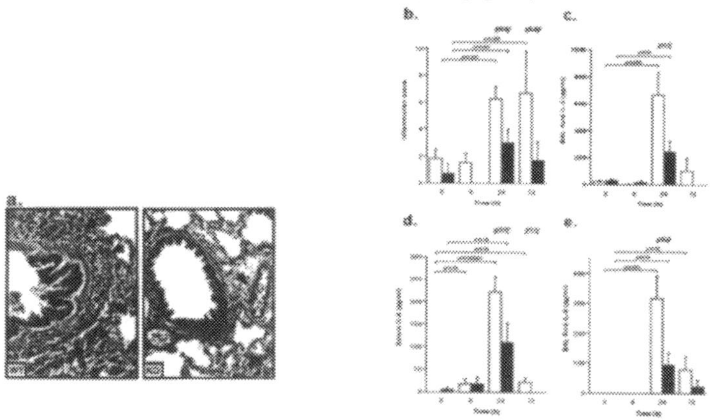

Figure 2. OVA-induced pulmonary inflammation in WT and KO mice. a: Photomicrographs of lungs of WT and KO mice 24 hours following challenge with the antigen (200 X). b: Inflammation score following histological evaluation of lungs from WT (Open bars) and KO (filled bars) mice. c-e: BAL fluid IL-4 (c), serum IL-5 (d) and BAL fluid IL-5 (e) levels as determined by ELISA in WT (Open bars) and KO mice (Filled bars) at various times following challenge with OVA. n= 4-19.

In conclusion, we have found that the EP_2 receptor is involved in PGE_2-mediated bronchodilation. Moreover, we propose that this receptor is also involved in the establishment of pulmonary inflammation following antigen challenge.

References

1. Pavord ID, Tattersfield AE. Bronchoprotective role for endogenous prostaglandin E2 [see comments]. Lancet 1995;345:436-38.
2. Sestini P, Armetti L, Gambaro G, et al. Inhaled PGE2 prevents aspirin-induced bronchoconstriction and urinary LTE4 excretion in aspirin-sensitive asthma. Am J Respir Crit Care Med 1996;153:572-75.
3. Gauvreau GM, Watson RM, O'Byrne PM. Protective effects of inhaled PGE2 on allergen-induced airway responses and airway inflammation. Am J Respir Crit Care Med 1999;159:31-36.
4. Pavord ID, Wong CS, Williams J, Tattersfield AE. Effect of inhaled prostaglandin E2 on allergen-induced asthma. Am Rev Respir Dis 1993;148:87-90.
5. Pavord ID, Wisniewski A, Mathur R, Wahedna I, Knox AJ, Tattersfield AE. Effect of inhaled prostaglandin E2 on bronchial reactivity to sodium metabisulphite and methacholine in patients with asthma. Thorax 1991;46:633-37.
6. Abramovitz M, Adam M, Boie Y, et al. Human prostanoid receptors: cloning and characterization. Adv Prostaglandin Thromboxane Leukot Res 1995;23:499-504.
7. Narumiya S. Molecular diversity of prostanoid receptors; subtypes and isoforms of prostaglandin E receptor. Adv Exp Med Biol 1997;400A:207-13.
8. Kennedy CR, Zhang Y, Brandon S, et al. Salt-sensitive hypertension and reduced fertility in mice lacking the prostaglandin EP2 receptor. Nat Med 1999;5:217-20.

INDUCTION OF PROSTAGLANDIN I_2 RECEPTOR IN MURINE OSTEOBLASTIC CELL

Kei Yamamoto, Jining Wang, Yukihiko Sugimoto, Atsushi Ichikawa and Shozo Yamamoto

Introduction

Prostanoids are important paracrine bioregulators for many animal tissues in physiological and pathological conditions. They are synthesized in the cyclooxygenase pathway, and released out of cells immediately. There are five basic types of prostanoid receptor linked to specific G proteins and these receptors show various biological activities [1]. MC3T3-E1 cloned from newborn mouse calvaria is an osteogenic cell line, which differentiated into an osteoblast [2]. Earlier we demonstrated production of PGE_2 as a major arachidonate metabolite in this cell line [3], and our later studies with MC3T3-E1 cells have shown the transcriptional regulation by TNFα of cyclooxygenase-2 induction [4]. In consideration of a possible function of the produced PGE_2 as an autocrine mediator, we attempted to detect prostanoid receptors in the plasma membrane of MC3T3-E1 cells. Previous papers reported the presence in this cell line of EP1, EP2, and EP4 [5], and FP [6], but their dynamic behaviors remained unclarified. In the present work, we will discuss an induction of IP rather than its constitutive expression in MC3T3-E1 cells [7].

Results and Discussion

When MC3T3-E1 cells were incubated with various concentrations of TNFα for 9 hours, there was a dose-dependent increase in the amount of IP mRNA. When the cells were incubated with TNFα, IP mRNA level was scarcely detectable at first, and started to increase after a lag time of about 3 hours reaching about 8-fold higher level at 24 hours (Figure 1A). Upon the addition of TNFα, cyclooxygenase-2 mRNA increased and reached a maximum faster than IP mRNA. In relation of the IP induction with IP function, we performed binding assay with TNFα-stimulated MC3T3-E1 cells. The cells were incubated in the presence or absence of TNFα for 24 hours, and the cell membranes were prepared. [^{3}H]-Labeled iloprost was utilized as a stable IP agonist, and its binding to the membrane protein was determined. Specific binding was scarcely detectable without TNFα. However, TNFα-stimulated cells showed iloprost binding increasing in a dose-dependent manner (Figure 1B). Furthermore, we examined the displacement of iloprost binding to the cell membrane by other agonists. [^{3}H]-Labeled iloprost at 40 nM was incubated with the

B. Samuelsson et al. (eds.), Advances in Prostaglandin and Leukotriene Research, 103–106.

membranes in the presence of PGE_1, PGE_2 or cicaprost each at 20 μM. The specific iloprost binding was displaced by 78% upon the addition of PGE_1, 39% with PGE_2, and 84% with cicaprost. Similar results were obtained by the positive control experiments with CHO cells expressing IP. Furthermore, we screened a mouse genomic library to clone the 5'-flanking region of IP gene. The promoter region contains various putative response elements such as SP1, E-BOX, C/EBPβ, and NFκB. A DNA fragment of the IP promoter region was prepared and inserted into the luciferase plasmid pGL3. MC3T3-E1 cells were transfected with the pGL3 plasmid containing IP gene. After 3 days, the cells were treated with TNFα at various concentrations and subjected to the luciferase analysis. The luciferase activity increased depending on the concentration of TNFα. The cells were incubated with TNFα for various time intervals. After a lag time of about 6 hours, the luciferase activity started to increase and reached a 5-fold higher level at 24 hours in comparison with the activity in the absence of TNFα (Figure 1C). These data indicated that the IP was increased in the TNFα-treated MC3T3-E1 cells.

It is well known that anti-inflammatory glucocorticoid suppresses the rapid and transient expression of cyclooxygenase-2. When the cells were incubated with TNFα in the presence of various concentrations of dexamethasone for 9 hours, the IP mRNA was suppressed by dexamethasone in a dose-dependent manner with an IC_{50} of about 50 nM. TNFα-dependent increase of IP promoter activity was also suppressed by dexamethasone dose-dependently. Thus, dexamethasone suppressed the IP induction at the step of transcription.

Then, we attempted to find out the mechanism of IP induction. The cells were incubated with TNFα for 9 hours. As examined by Northern blotting, cycloheximide as a translation inhibitor markedly reduced the TNFα-induced increase in IP mRNA. Therefore, it is likely that the IP expression is dependent on *de novo* synthesis of a protein or proteins. The cyclooxygenase inhibitors such as indomethacin and NS398 also inhibited the IP mRNA induction by about 35% and 55%, respectively. Then, TNFα-dependent increase of IP promoter activity was partially inhibited by indomethacin and NS398 dose dependently. Furthermore, we tested the effects of several prostanoids on the IP mRNA induction. The addition of PGE_1, PGE_2, $PGF_{2\alpha}$, or iloprost increased the IP mRNA level by 3- to10-fold. As for the production of prostanoids in this cell line, it was confirmed that PGE_2 was essentially a sole arachidonate metabolite in the TNFα-stimulated cells by thin-layer chromatography and high performance liquid chromatography. Since these cyclooxygenase inhibitors suppressed the IP induction to a considerable extent, and PGE_2 increased the IP mRNA level, the cyclooxygenase-2 induction which leads to the production of PGE_2 may be functionally coupled to the IP induction.

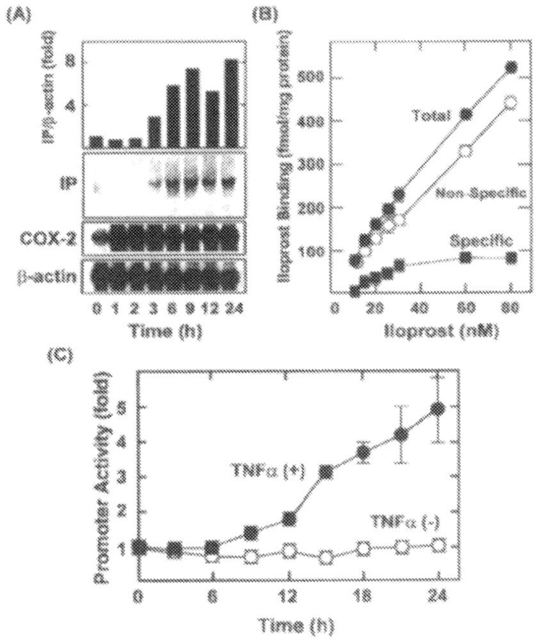

Figure 1. Induction of IP by TNFα in MC3T3-E1 cells. (A) Time-dependent effect of TNFα on IP mRNA. The cells (2.2 x 10^6 cells) were incubated with 10 ng/ml TNFα for indicated time periods. Total RNA (20 μg) was subjected to Northern blotting for IP, cyclooxygenase-2 (COX-2), and β-actin mRNA. The relative amount of IP mRNA was corrected for β-actin mRNA. (B) [^{3}H]Iloprost binding to cell membrane. The cells (1.0 x 10^8 cells) were incubated with 20 ng/ml TNFα for 24 hours, and the cell membranes were prepared. The membranes (100 μg protein) were incubated with [^{3}H]iloprost at various concentrations for 1 hour at 30 °C. The specific binding (closed square) was determined by subtraction of the nonspecific binding (open circle) from the total binding (closed circle). (C) Transfection analysis of IP gene. Promoter region of mouse IP gene was ligated to a luciferase plasmid pGL3. Confluent cells (3.1 x 10^5 cells) were co-transfected with luciferase plasmid including mouse IP gene and β-galactosidase plasmid. The cells were incubated with 20 ng/ml TNFα for indicated time intervals. The luciferase activity was assayed, and the results were normalized with the β-galactosidase activity.

References

1. Narumiya S, Sugimoto Y, Ushikubi F. Prostanoid receptors: structures, properties, and functions. Physiol Rev 1999;79:1193-1226.
2. Kodama H, Amagai Y, Sudo H, Kasai S, Yamamoto S. Establishment of a clonal osteogenic cell line from newborn mouse calvaria. Jpn J Oral Biol 1981;23:899-901.

3. Yokota K, Kusaka M, Ohshima T, et al. Stimulation of prostaglandin E_2 synthesis in cloned osteoblastic cells of mouse (MC3T3-E1) by epidermal growth factor. J Biol Chem 1986;261:15410-15.
4. Yamamoto K, Arakawa T, Ueda N, Yamamoto S. Transcriptional roles of nuclear factor κB and nuclear factor-interleukin-6 in the tumor necrosis factor α-dependent induction of cyclooxygebase-2 in MC3T3-E1 cells. J Biol Chem 1995;270:31315-20.
5. Suda M, Tanaka K, Natsui K, et al. Prostaglandin E receptor subtypes in mouse osteoblastic cell line. Endocrinology 1996;137:1698-1705.
6. Nemoto K, Bernecker PM, Pilbeam CC, Raisz LG. Expression and regulation of prostaglandin F receptor mRNA in rodent osteoblastic cells. Prostaglandins 1995;50:349-58.
7. Wang J, Yamamoto K, Sugimoto Y, Ichikawa A, Yamamoto S. Induction of prostaglandin I_2 receptor by tumor necrosis factor α in osteoblastic MC3T3-E1 cells. Biochim Biophys Acta 1999;1441:69-76.

BLT2, A SECOND LEUKOTRIENE B_4 RECEPTOR

Takehiko Yokomizo, Kazuhiko Kato, Kan Terawaki, Takashi Izumi, and Takao Shimizu

Introduction

Leukotriene B_4 (LTB_4) is a potent chemoattractant and activator for granulocytes and macrophages [1-3]. The high-affinity receptor for LTB_4 (BLT1) was cloned and shown to be a G-protein-coupled receptor (GPCR) with low structural homology to the other known GPCRs [4]. Here we report a second LTB_4 receptor, BLT2, found in a gene cluster with BLT1 [5,6]. BLT2 is the most homologous GPCR to BLT1, and is identified as a low-affinity receptor for LTB_4.

Identification of BLT2 in a Gene Cluster of BLT1

After the initial cloning of human BLT1, we and others cloned BLT1 from various animals [7-11]. BLT1 expression is limited only in leukocytes in these animals, and enhanced BLT1 mRNA was observed in eosinophils in IL-5 transgenic mice, and in stimulated peritoneal macrophages in rat and mouse [8,11]. Thus, transcription of BLT1 is regulated in tissue-specific and inducible manners. To clarify the mechanism how BLT1 is regulated, we isolated human and mouse genes for BLT1, and analyzed the transcriptional regulation of BLT1 [5]. During the course of the study, we identified a putative ORF for a novel GPCR similar to BLT1 at the 5 portion of BLT1 gene. As this putative GPCR was revealed to be another receptor for LTB_4, we named it BLT2. BLT2 is the most homologous GPCR to BLT1 with the amino acid identity of 45.2% (Figure 1), which is much higher than those to orphan GPCRs, GPR25 and CRTH2 [12, 13]. The amino acid identity between human and mouse BLT2 is 92.7% which is higher than that of BLT1 (~80%) between two species, showing that BLT2 has been well conserved evolutionally. Northern blotting in human tissues revealed the ubiquitous expression of BLT2, with the highest expression in spleen, followed by liver, ovary, and leukocytes. The tissue distribution of BLT2 is quite different from BLT1, suggesting that these two receptors may have distinct roles *in vivo*.

BLT2 is a Low-affinity Rreceptor for LTB_4

When expressed in HEK-293 cells, BLT2 showed a specific and saturable binding for [^{3}H]

B. Samuelsson et al. (eds.), Advances in Prostaglandin and Leukotriene Research, 107–110.

LTB_4 with the *Kd* value of 23 nM, a value higher than that of BLT1 (1.1 nM). Two BLT antagonists were tested for their inhibition of [^{3}H] LTB_4 binding to BLT1 and BLT2. ONO-4057 was able to antagonize both BLT1 and BLT2, while U75302 failed to antagonize BLT2, showing that these two receptors are pharmacologically distinct. The low affinity receptor for LTB_4 was reported in spleen membrane [14,15], and BLT2 is expressed highest in spleen, suggesting that BLT2 is the reported low affinity receptor for LTB_4.

```
hBLT1     1 M-----NTTSSAAPPSLGVEFISLLAIILLSVALAVGLPGNSFVVWSILKRMQKRS--VT    53
hBLT2     1 MSVCYRPPGNETLL-.WKTSRATGT.FL..AAL.--.....G.....LAGWRPA.GRPLA   57
                                              I
hBLT1    54 ALMVLNLALADLAVLLTAPFFLHFLAQGTWSFGLAGCRLCHYVCGVSMYASVLLITAMSL  113
hBLT2    58 .TL..H.....G....LT.L.VA..TRQA.PL.Q...KAVY...AL........TGLL..  117
                     II                                                 III
hBLT1   114 DRSLAVARPFVSQKLRTKAMARRVLAGIWVLSFLLATPVLAYRTVVPWKTNMS-LCFPRY  172
hBLT2   118 Q.C...T...LAPR..SP.L...L.LAV.LAAL...V.AAV..H--L.RDRVCQ..H.S-  174
                                                      IV
hBLT1   173 PSEGHRAFHLIFEAVTGFLLPF-LAVVASYSDI-GRRLQARRFRRSR---RTGRLVVLII  227
hBLT2   175 .V--.A.A..SL.TL.A.V...G.-MLGC..VTLA-..RGA.WGSG.HGA.V.....SA.V 230
                                        V
hBLT1   228 LTFAAFWLPYHVVNLAEAGRALAGQAAGLGLVGKRLSLARNVLIALAFLSSSVNPVLYAC  287
hBLT2   231 .A.GLL.A...A...LQ.VA...PPEGA.AKL.GAGQA..AGTT....F.........VF  290
                VI                                                     VII
hBLT1   288 AGGGLLRSAGVGFVAKLLEGTGSEASSTRRG-GSLGQTARSGPAALEPGPSESLTASSPL  346
hBLT2   291 TA.D..PR..PR.LTR.F..S.EARGGG.SRE.TMELRTTPQLKVVGQ.RGNGDPGGG-M  349

hBLT1   347 -KLNE-LN- 352
hBLT2   350 E.DGPEWDL 358
```

Figure 1. Primary structures of human BLT1 and BLT2. Identical amino acids are shown by dots. Putative transmembrane domains are shown by underlines.

Signaling from BLT2

We established CHO cells expressing stably BLT1 or BLT2, and compared the intracellular signaling using these cells. BLT2 transduces signals toward adenylyl cyclase inhibition, calcium mobilization, and chemotaxis like BLT1 except for that higher concentration of LTB_4 is required. BLT2-mediated inhibition of adenylyl cyclase was not inhibited by pertussis toxin pretreatment (PTX), showing that this receptor couples to Gz-like G-protein which lacks the Cys residue that can be ADP-ribosylated by PTX. Table 1 summarizes the differences between BLT1 and BLT2.

We should note that ORF of BLT2 overlaps the promoter of BLT1 in human gene [5]. This is, to our knowledge, the first example of so called romoter in ORF in mammals. There are many questions to be answered. Why these two receptors are clustered in such a small portion of the gene? Are there any relationship between the transcriptional regulation of BLT1 and translational regulation of BLT2? What is the physiological and pathophysiological role of BLT1 and BLT2?

Table 1. Comparison of BLT1 and BLT2

	BLT1	BLT2
Amino acids (human)	352	358
Kd in HEK-293 cells	1.1 nM	23 nM
Expression	Leukocytes	Spleen
	>>Thymus, Spleen	>Liver, Ovary, Leukocytes
PLC activation	G16, Gi2	Gq-like, Gi-like
Adenylyl cyclase inhibition	Gi-like	Gz-like
Chemotaxis	Gi-like	Gi-like

Acknowledgments

This work was supported in part by, Grants-in-Aid from the Ministry of Education, Science, Sports, and Culture and Human Science Foundation, and grants from the Yamanouchi Foundation for Metabolic Disorders, the Uehara Memorial Foundation, and the Cell Science Research Foundation.

References

1. Serhan CN, Haeggstrom JZ, Leslie CC. Lipid mediator networks in cell signaling: update and impact of cytokines. Faseb J 1996;10:1147-58.
2. Samuelsson B, Dahlen SE, Lindgren JA, Rouzer CA, Serhan CN. Leukotrienes and lipoxins: structures, biosynthesis, and biological effects. Science 1987;237:1171-76.
3. Shimizu T, Wolfe LS. Arachidonic acid cascade and signal transduction. J Neurochem 1990;55:1-15.
4. Yokomizo T, Izumi T, Chang K, Takuwa Y, Shimizu T. A G-protein-coupled receptor for leukotriene B^4 that mediates chemotaxis. Nature 1997;387:620-24.
5. Kato K, Yokomizo T, Izumi T, Shimizu T. Cell-specific transcriptional regulation of human leukotriene B_4 receptor gene. J Exp Med 2000;192:413-20.
6. Yokomizo T, Kato K, Terawaki K, Izumi T, Shimizu T. A second leukotriene B_4 receptor, BLT2: a new therapeutic target in inflammation and immunological disorders. J Exp Med 2000;192:421-31.
7. Boie Y, Stocco R, Sawyer N, et al. Characterization of the cloned guinea pig leukotriene B_4 receptor: Comparison to its human orthologue. Eur J Pharmacol 1999;380:203-13.
8. Huang WW, Garcia-Zepeda EA, Sauty A, Oettgen HC, Rothenberg ME, Luster AD. Molecular and biological characterization of the murine leukotriene B_4 receptor expressed on eosinophils. J Exp Med 1998;188:1063-74.
9. Martin V, Ronde P, Unett D, et al. Leukotriene binding, signaling, and analysis of HIV coreceptor function in mouse and human leukotriene B_4 receptor-transfected cells. J Biol Chem 1999;274:8597-603.

10. Masuda K, Yokomizo T, Izumi T, Shimizu T. cDNA cloning and characterization of guinea-pig leukotriene B_4 receptor. Biochem J 1999;342:79-85.
11. Toda A, Yokomizo T, Masuda K, Nakao A, Izumi T, Shimizu T. Cloning and characterization of rat leukotriene B_4 receptor. Biochem Biophys Res Commun 1999;262: 806-12.
12. Jung BP, Nguyen T, Kolakowski LF, Jr., et al. Discovery of a novel human G protein-coupled receptor gene (GPR25) located on chromosome 1. Biochem Biophys Res Commun 1997;230:69-72.
13. Nagata K, Tanaka K, Ogawa K, et al. Selective expression of a novel surface molecule by human Th2 cells in vivo. J Immunol 1999;162:1278-86.
14. Showell HJ, Breslow R, Conklyn MJ, Hingorani GP, Koch K. Characterization of the pharmacological profile of the potent LTB_4 antagonist CP-105,696 on murine LTB_4 receptors in vitro. Br J Pharmacol 1996;117:1127-32.
15. Miki I, Watanabe T, Nakamura M, et al. Solubilization and characterization of leukotriene B_4 receptor-GTP binding protein complex from porcine spleen. Biochem Biophys Res Commun 1990;166:342-48.

Upregulation of Cysteinyl-Leukotriene 1 Receptor Expression by Interleukin-5 in Human Leukocytes

Marek Rola-Pleszczynski, Maryse Thivierge, and Jana Stankova

Introduction

The cysteinyl-leukotrienes (cysLTs), LTC_4, LTD_4, and LTE_4 are potent lipid mediators implicated mainly in acute bronchoconstriction and chronic airway inflammation in asthma [1]. Blood eosinophils from asthmatic patients synthesize increased amounts of cys-LTs. Cys-LTs, in turn, have *in vitro* chemoattractant activity for human eosinophils. LTD4-induced eosinophil infiltration of the airway could be blocked by the cysLT antagonist MK-571 [2].

Recently, Lynch et al. reported the successful cloning and expression of a high-affinity cell-surface human LTD_4 receptor ($CysLT_1R$) [3].

IL-5 is one of the most important regulators of eosinophil functions including chemotaxis, degranulation, adhesion, and cytotoxicity for parasites. IL-5 is synthesized by Th2 lymphocytes and mast cells and is detectable in the plasma of symptomatic asthmatics [4].

HL-60 cells differentiated toward the eosinophil phenotype (HL-60/eos) present several characteristics of mature eosinophils, including $CysLT_1R$ expression [5]. In the present study, we set out to determine whether the cytokine IL-5 could modulate the expression of $CysLT_1R$ on HL-60/eos cells as well as on human peripheral blood leukocytes. We found IL-5 to be a potent upregulator of $CysLT_1R$ gene expression in leukocytes.

Materials and Methods

Differentiation toward the eosinophil phenotype (HL-60/eos) was induced as previously reported [5]. Peripheral blood monocytes and neutrophils were isolated by gradient centrifugation on Hypaque-Ficoll.

For semi-quantitative RT-PCR analysis, total RNA was extracted from cells and converted to cDNA by the reverse transcriptase enzyme reaction (AMV transcriptase-reverse, Promega, Madison, WI). $CysLT_1R$ cDNA was amplified by PCR with primers derived from the published cDNA sequence for $CysLT_1R$ [3]. Samples were subjected to parallel amplification of the constitutively expressed, housekeeping gene, GAPDH.

B. Samuelsson et al. (eds.), Advances in Prostaglandin and Leukotriene Research, 111–113.

The expression of $CysLT_1R$ protein on cells was assessed using a polyclonal anti-$CysLT_1R$ Ab (Dr J. Johnson, Cayman Chemical, Ann Arbor, MI). Immunofluorescence analysis was performed on a FACScan flow cytometer (Becton-Dickinson, San Jose, CA).

Results and Discussion

To assess the effect of IL-5 on $CysLT_1R$ expression, we analyzed $CysLT_1R$ mRNA expression by RT-PCR in HL-60/eos cells, as well as in peripheral blood neutrophils and monocytes. As illustrated in Figure 1, IL-5 induced a marked augmentation of $CysLT_1R$ mRNA expression in all three leukocyte populations. The effect of IL-5 on $CysLT_1R$ was maintained up to 24 hours (data not illustrated).

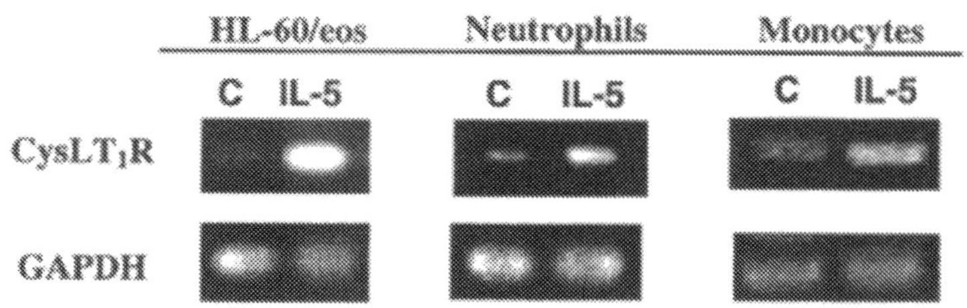

Figure 1. Upregulation of $CysLT_1R$ mRNA expression by IL-5.

To assess whether transcription of $CysLT_1R$ mRNA was associated with an augmented expression of $CysLT_1R$ protein at the cell surface, flow cytometry studies were performed using a polyclonal antibody directed against the C-terminal portion of $CysLT_1R$. For these experiments, cells were cultured for 24 hours in the absence or presence of 10 ng/ml IL-5, before labelling with the antibody. As illustrated in Figure 2, IL-5 treatment of monocytes induced a markedly augmented expression of $CysLT_1R$ protein, as compared to untreated cells.

Our results show, for the first time, that the cytokine IL-5 rapidly upregulates $CysLT_1R$ mRNA expression and that this induction is associated with enhanced $CysLT_1R$ protein expression in leukocytes. These newly induced receptors are functional since pretreatment of cells with IL-5 is associated with an augmented responsiveness to the ligand LTD_4 in terms of intracellular calcium mobilization and chemotaxis [Thivierge M et al. J Immunol, submitted].

IL-5 (10ng/ml)

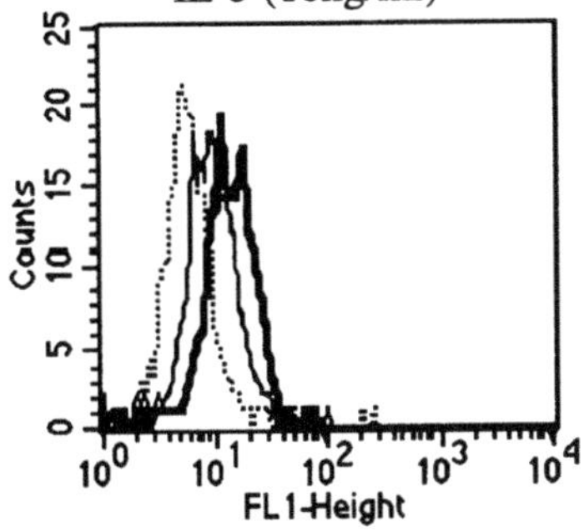

Figure 2. Upregulation of $CysLT_1R$ expression on human monocytes by IL-5. Human peripheral blood monocytes were treated for 24 hours with IL-5 or medium and $CysLT_1R$ expression was assessed by FACS. Dotted line represents nonpertinent Ab and FITC control. Fine line represents unstimulated cells, thick line represents IL-5-stimulated cells, both stained with anti-$CysLT_1R$ Ab and FITC-conjugated goat anti-rabbit secondary Ab.

References

1. Dahlen SE, Hedqvist P, Hammarstrom S, Samuelsson B. Leukotrienes are potent constrictors of human bronchi. Nature 1980;288:484-86.
2. Chan C.C., McKee K, Tagari P, Chee P, Ford-Hutchinson A. Eosinophil-eicosanoid interactions:inhibition of eosinophil chemotaxis in vivo by a LTD4 receptor antagonist. Eur J Pharmacol 1990;191:273-80.
3. Lynch KR, O'Neill GP, Liu Q, et al. Characterization of the human cysteinyl leukotriene CysLT1 receptor. Nature 1999;399:789-93.
4. Corrigan CJ, Haczku A, Gemou-Engesaeth V, et al. CD4 T-lymphocyte activation in asthma is accompanied by increased serum concentrations of interleukin-5. Effect of glucocorticoid therapy. Am Rev Respir Dis 1993;147:540-47.
5. Patry C, Muller E, Laporte J, Rola-Pleszczynski M, Sirois P, de Brum-Fernandes AJ. Leukotriene receptors in HL-60 cells differentiated into eosinophils, monocytes and neutrophils. Prostaglandins Leukot Essent Fatty Acids 1996;54:361-70.

Physiological and Pathophysiological Roles of Prostanoids; Lessons from Receptor-Knockout Mice for Clinical Applications

Shuh Narumiya

Introduction

Prostanoids including prostaglandins (PGs) and thromboxanes (TXs) exert their actions by acting on eight types and subtypes of prostanoid receptors [1]. They are the PGD receptor, DP; four subtypes of PGE receptors: EP_1, EP_2, EP_3, and EP_4; the PGF receptor, FP; the PGI receptor, IP; and the TXA receptor, TP. These receptors are encoded by different genes, and all of them are rhodopsin-type, G-protein-coupled receptors with seven transmembrane domains. Dependent on changes in intracellular second messengers, these receptors can be grouped into three. One group consists of TP, FP, and EP_1, which couple to a rise in intracellular free calcium ion. The second group includes DP, EP_2, EP_4, and IP, which induce a rise in cAMP in cells. The last receptor, EP_3, couples to a decrease in cAMP level. These prostanoid receptors are not distributed ubiquitously and evenly in the body; each receptor has its own unique distribution pattern. They are also subject to induction or suppression in response to various physiological and pathophysiological stimuli. A variety of physiological actions of prostanoids are thus exerted by a molecular diversity of prostanoid receptors mediating different signal transductions in the cell and distributed characteristically in the body. It is, however, not necessarily clear which type of prostanoid receptors is responsible for a prostanoid action in a particular process. Neither it is known how much prostanoids contribute to various physiological and pathophysiological processes. To address these issues, we have individually disrupted genes for all eight types of prostanoid receptors, and analyzed mice deficient in each receptor in various pathological contexts. This article summarizes our analyses on the roles of prostanoids and their receptors in pain transmission, fever generation, and allergic asthma. Development of new PG analogs based on the cloned receptors are also discussed in conjunction of knockout mice studies.

PGI_2, IP Receptor, and Pain Sensation

Pain is sensed at the periphery of the primary afferent and a signal is transmitted to dorsal horn of the spinal cord. It is then relayed to secondary neurons that extend their axons up to the brain stem and thalamus. Classically, prostanoids are believed to sensitize the

B. Samuelsson et al. (eds.), Advances in Prostaglandin and Leukotriene Research, 115–120.

afferent in the periphery to cause hyperalgesia. The primary afferents mediating pain have their cell bodies in the dorsal root ganglion (DRG). To identify prostanoid receptors involved in pain sensation, we carried out *in situ* hybridization analysis of prostanoid receptor mRNA expressed in neurons in DRG [2,3]. On quantification, about 29% of total DRG neurons express EP_1, 21% EP_4, 50% EP_3, and 40% IP. In addition, about two thirds of DRG neurons expressing PPTA, a precursor of substance P, co-express IP mRNA, indicating that neurons expressing IP are involved in mediation of pain. We examined the role of IP in pain mediation by disrupting the IP gene and subjecting mice deficient in IP to acetic acid writhing [4]. When wild type mice were subjected to this test, they showed about 40 times writhing responses in 30 minutes. Indomethacin treatment decreased the number to less than 10 times, showing PG-dependent sensitization of pain sensation. When IP-deficient mice were subjected, they showed only less than 10 times responses without indomethacin treatment, and indomethacin did not decrease the response time further. We also found that a PGI_2 analog injected i.p also caused pain response, albeit weak, in wild type but not in IP-deficeint mice. These results taken together suggest that hyperalgesia to acetic acid is mediated by the action of the PGI_2 and IP receptor pathway in the periphery. Then, the next question is whether IP mediates every kind of pain sensation or other receptors expressed in DRG neurons are also involved in pain sensation in different contexts. Recent experiments using intrathecal injection of COX inhibitors suggest that prostanoids work not only in the periphery to cause hyperalgesia but also in the dorsal horn to facilitate transmission of pain sensation [5]. The latter type of actions are related to allodynia, a painful response to a normally nonpainful stimulus, often seen in pain response to neuropathy. Involvement of EP receptors in this and other types of pain responses are currently under investigation.

PGE_2, EP_3 Receptor, and Fever Generation

Sickness causes various symptoms in animals, including humans, such as fever, ACTH release, loss of appetite, decreased locomotion, and increased slow wave sleep. Among them, involvement of PGs in fever generation has been suggested by antipyretic action of aspirin-like drugs. However, identity of a prostanoid and a prostanoid receptor involved in febrile response has remained elusive. PGE_2, when injected intracerebroventricularly (i.c.v.), causes fever in mice. We therefore used mice deficient in each of four subtypes of PGE receptor and examined their response to i.c.v. injection of PGE_2 [6]. PGE_2 generated fever that had a peak rise of about 2 °C at 20 minutes after injection in wild type, EP_1-, EP_2- and EP_4-deficient mice. However, no fever generation was observed in EP_3-deficient mice, suggesting that febrile response to i.c.v. injected PGE_2 is mediated by EP_3 receptor in the brain. We wondered if the same mechanism operates in fever generation in response to various pyrogens. To clarify this point, we injected mice with either an endogenous pyrogen, interleukin-1β (IL-1β) or an exogenous pyrogen, lipopolysaccharide (LPS). Intravenous injection of IL-1 induced fever in wild type mice, that again peaked at 20 minutes after the injection and subsided thereafter. Mice deficient in EP_1 or EP_2 showed similar febrile responses. On the other hand, no fever was observed in EP_3-deficient homozygous mice. Loss of fever generation was also noted in EP_3 KO mice upon challenge with LPS. These results clearly show that EP_3 links pyrogen signals to generation of fever. We next wondered where in the brain this linking event occurs. Blood-borne signals such as cytokines and LPS are believed to act on areas of the brain where the blood brain barrier is poor. One of such areas is organum vasculosum lamina terminalis (OVLT), situated around the third ventricle. Autoradiograms of *in situ* hydridization for EP_3 mRNA expression in the brain [2] showed that, although the EP_3 signal is not found in OVLT itself, it is present as numerous dots in the surrounding area. These dots correspond to neurons in this area. These results thus suggest that pyrogens such as cytokines and LPS act in OVLT to generate PGE_2, that then acts on EP_3 in neighboring neurons to trigger the febrile response.

PGD_2, DP Receptor, and Allergic Asthma

Allergic asthma is caused by activation of IgE bound on mast cells with specific allergen, and shows airway inflammation, airway obstruction, and airway hypersensitivity [7]. The airway inflammation associated with asthma is characterized by infiltration of TH2 lymphocytes and eosinophilic leukocytes. TH2 cytokines released from the infiltrated lymphocytes such as interleukin (IL)-4, IL-5, and IL-13 are believed to play critical roles in initiation and progression of asthma. The airway obstruction occurs as a result of both contraction of airway smooth muscle and mucus secretion. There is plenty of evidence to suggest the involvement of PGD_2 in allergic asthma. PGD_2 is the major prostaglandin formed by mast cells upon allergen challenge [8], and is released in large amounts during asthmatic attack in human subjects [9]. However, whether PGD_2 acts as a mediator of allergic asthma or down-regulates this process remains unknown. To clarify this issue, we generated mice deficient in PGD receptor [10]. The mouse DP gene was disrupted by insertion of a neomycin resistance gene into the first coding exon. DP-deficient mice and their wild type littermates were sensitized to ovalbumin (OVA) by intraperitoneal injection of OVA on day 0 and day 12. On day 22, 26, and 30, these mice were challenged with aerosolized OVA for 30 minutes, and IgE level, cell infiltration, cytokine production, mucus formation, and the airway hypersensitivity were examined. The total IgE and OVA-specific IgE concentrations were markedly increased in response to injection of OVA. There was no significant difference in these IgE levels between wild type and DP-deficient mice. In spite of similar IgE levels, mice deficient in DP showed marked reduction in cell infiltration in both lung and bronchoalveolar lavage fluid (BALF). TH2 cytokines, IL-4, IL-5, and IL-13 that were released extensively to the BALF by antigen challenge in wild type animals were also markedly reduced in DP-deficient mice. The airway of DP-deficient mice showed reduced PAS staining in the airway epithelial cells, indicating reduced mucus secretion, and little hypersensitivity to acetylcholine infused intravenously. Thus, in DP-deficient mice, all of the three major symptoms of allergic asthma, the TH2-dependent airway inflammation, the airway obstruction, and the airway hypersensitivity, were suppressed, suggesting a critical role of PGD_2-DP receptor signaling in triggering allergic reactions. Then, where in the allergic lung does this signaling occur? Immunohistochemistry using specific anti-DP antibody revealed that DP receptor is extensively induced in the lung after the antigen challenge to the airway and immunoelectron microscopy revealed that it is expressed in epithelial cells of the bronchioles and alveoli. Recently, evidence has accumulated to indicate that these epithelial cells are not a simple barrier but contain various chemokines and cytokines possibly working in allergy [7].

Based on these results, we currently hypothesize the PGD_2 action in allergic asthma as follows. PGD_2 is released upon antigen challenge and acts on DP receptor in airway epithelial cells to stimulate synthesis and/or release of these cytokines, which in turn trigger allergic inflammation and subsequent asthmatic events. In addition to the lung, PGD_2 is produced in various other allergic disorders, such as allergic rhinitis and atopic dermatitis, suggesting that it may also play an important role in these allergic states. The DP receptor may thus represent a new therapeutic target for allergic reactions.

Cloned Prostanoid Receptors as a Platform of New Drug Development

One of the advantages to have a family of cloned receptors is to use them for assessment of compounds for development of new drugs. We stably expressed each of the cloned prostanoid receptors in Chinese hamster ovary cells, and first used them to evaluate binding properties of a series of PGs and conventional PG analogs used in pharmacological analysis [11]. Table 1 shows an example of such analysis. Compounds shown are two prostacyclin analogs, two thromboxane agonists, and PGE_2 and 17-phenyl-PGE_2. As shown here, there is no absolute selectivity in any of these compounds. For example, iloprost and carbacyclin indeed bind to IP with good affinities, but they also bind to EP_3 with similar or better affinities. Iloprost binds to EP_1 as well. 17-phenyl-PGE_2 has been used as an EP_1 agonist, but its binding is not limited to EP_1. It binds to EP_3 with a better affinity and to FP with a reasonably good affinity.

Conventional PG analogs thus show considerable cross-reactivities. This is probably due to their screening and characterization on native tissues and cells, because most tissues and cells express more than one type of prostanoid receptors. By using a panel of cloned prostanoid receptors, Ono Pharmaceuticals in collaboration with us has successfully developed highly selective agonists and antagonists to each type of the prostanoid receptors (Table 2) [12,13]. For example, an EP1 antagonist, ONO-8713, shows a Ki value of 0.3 nM to EP1, and the next closest Ki is 1000 nM for EP3. So, there is more than 3,000 times difference in binding affinity between EP1 and other PG receptors to this compound. This is also true for recently developed EP agonists. For example, AE1-329 shows a Ki value of 10 nM to EP4, but Ki values of more than 1200 nM to other EP subtypes. Development of other selective EP1 antagonists and a highly selective DP agonist by a similar strategy has also been reported by the Merck-Frosst group [14,15]. A DP agonist, L-644,698 showed the Ki of 0.9 nM for human DP, which was about 300 times lower than that for EP2 and more than 4,000 times lower than those for other types of receptors [15].

Table 1. Binding Affinities of Several PG analogs to Cloned Prostanoid Receptors

Compounds	DP	EP_1	EP_2	EP_3	EP_4	FP	IP	TP
Iloprost	>10000	21	1600	22	2300	>10000	10	>10000
carbacyclin	>10000	>10000	1600	31	2300	1200	110	>10000
I-BOP	>10000	>10000	220	100	>10000	100	>10000	1
STA2	1600	>10000	220	23	>10000	97	>10000	14
PGE_2	>10000	2	12	1	2	100	>10000	>10000
17-Phenyl-PGE_2	>10000	14	>10000	4	1000	60	>10000	>10000

Ki values for radioligand binding assay for respective receptor are shown in nM.

Table 2. Binding Affinities of Novel EP Agonists and Antagonists to Cloned Prostanoid Receptors

Compounds	DP	EP_1	EP_2	EP_3	EP_4	FP	IP	TP
EP_1 antagonists								
ONO-8711	1800	2	5300	67	>10000	>10000	>10000	8
ONO-8713	>10000	0.3	3000	1000	>10000	1400	10000	10000
EP agonists								
ONO-DI-004		150	>10000	>10000	>10000			
ONO-AE1-259		>10000	3	>10000	>10000			
ONO-AE-248		>10000	3700	8	4200			
ONO-AE1-329		>10000	2100	1200	10			

Ki values for radioligand binding assay for respective receptor are shown in nM.

Conclusion

KO mouse studies have told us and are telling us a lot of lessons as to roles of prostanoids in various physiological and pathophysiological processes. Besides the findings described above, many other implicative phenotypes have been found in KO mice deficient in prostanoid receptors [1]. EP_1-deficient mice show reduced formation of precancerous lesions to a colon carcinogen, suggesting involvement of EP_1 in colon carcinogenesis. EP_2-KO mice show decreased fertilization due to the abortive expansion of the cumulus covering the oocyte. EP_4-KO mice show patent ductus arteriosus and decreased bone resorption. FP-deficient mice show loss of parturition due to abortive luteolysis. IP-KO show increased thrombotic tendency and increased inflammatory swelling, and TP-KO bleeding tendency and enhanced immune responses. These lessons should be combined with new highly selective agonists and antagonists screened on the panel of cloned prostanoid receptors, and be fully exploited to develop drugs to treat and cure diseased people.

References

1. Narumiya S, Sugimoto Y, Ushikubi F. Prostanoid receptors; structures, properties and functions. Physiol Rev 1999;79:1193-1226.
2. Sugimoto Y, Shigemoto R, Namba T, et al. Distribution of the messenger RNA for the prostaglandin E receptor subtype EP3 in the mouse nervous system. Neuroscience 1994; 62:919-28.
3. Oida H, Namba T, Sugimoto Y, et al. *In situ* hybridization studies of prostacyclin receptor mRNA expression in various mouse organs. Br J Pharmacol 1995;116:2828-37.
4. Murata T, Ushikubi F, Matsuoka T, et al. Altered pain perception and inflammatory responses in mice lacking prostacyclin receptor. Nature 1997;388:678-82.
5. Malmberg AB, Yaksh TL. Hyperalgesia mediated by spinal glutamate or substance P receptor blocked by spinal cyclooxygenase inhibition. Science 1992;257:1276-79.
6. Ushikubi F, Sugimoto Y, Murata T, et al. Impaired febrile response in mice lacking the prostaglandin E receptor subtype EP3. Nature 1998;395:281-84.
7. Holgate ST. The epidemic of allergy and asthma. Nature 1999;402(Suppl.):B2-B4.
8. Lewis RA, Soter NA, Diamond PT, Austen KF, Oates JA, Roberts LJ II. Prostaglandin D_2 generation after activation of rat and human mast cells with anti-IgE. J Immunol 1982; 129:1627-31.
9. O'Sullivan S. On the role of PGD_2 metabolites as markers of mast cell activation in asthma. Acta Physiol Scand Suppl 1999;644:1-74.
10. Matsuoka T, Hirata M, Tanaka H, et al. Prostaglandin D_2 as a mediator of allergic asthma. Science 2000;287:2013-17.
11. Kiriyama M, Ushikubi F, Kobayashi T, Narumiya S. Ligand binding specificities of the eight types and subtypes of the mouse prostanoid receptors expressed in Chinese hamster ovary cells. Br J Pharmacol 1997;122:217-24.
12. Watanabe K, Kawano T, Nakatsugu S, et al. Role of the prostaglandin E receptor subtype EP1 in colon carcinogenesis. Cancer Res 1999;59:5093-96.
13. Suzawa T, Miyaura C, Inada M, et al. The role of prostaglandin E receptor subtypes (EP1, EP2, EP3 and EP4) in bone resorption: an analysis using specific agonists for the resepctive EPs. Endocrinology 2000;141:1554-59.
14. Ruel R, Lacombe P, Abramovitz M, et al. New Class of biphenylene dibenzazocinones as potent ligands for the human EP1 prostanoid receptor. Bioorg Med Chem Lett 1999;9:2699-704.
15. Wright DH, Metters KM, Abramovitz M, Ford-Hutchinson AW. Characterization of the ecombinant human prostanoid DP receptor and identification of L-644,698, a novel selective DP agonist. Br J Pharmacol 1998;123:1317-24.

Multiple G-Protein Coupling of Chimeric Prostacyclin/prostaglandin D2 Receptors

Helen Wise and Kevin B.S. Chow

Introduction

The prostacyclin (IP) receptor couples to both G_s and G_q-proteins, with coupling to G_s being the dominant signaling pathway [1]. Modification of the C-terminal tail of the mouse IP receptor (mIP) with corresponding regions from the mouse prostaglandin D_2 receptor (mDP) results in a constitutively active mutant receptor (IP_{N-VII}/DP_C) which displays a 20-fold increase in potency for coupling to G_s [2]. However, when the 6th to 7th transmembrane domains of the mIP receptor are also replaced by the corresponding regions of the mDP receptor (IP_{N-V}/DP_{VI-C}), this gain in potency is lost. We have therefore used these chimeric IP/DP receptors to examine the differential coupling between G_s and G_q, and have concluded that the C-terminal tail of the mIP receptor cannot be entirely responsible for its coupling to G_q.

Methods

Chinese hamster ovary (CHO) cells were transiently transfected with cDNA using Lipofectamine liposome reagent and Opti-mem I reduced serum medium, using cDNA concentrations aimed at producing similar levels of receptor expression (approximately 100 fmol/mg protein) [2]. Cell responses were assayed in 12-well plates, 48-hour post-transfection, in HEPES-based buffer containing 3 μM indomethacin to inhibit any endogenous prostanoid production. [^{3}H]-cyclic AMP production from [^{3}H]-labelled cells was assayed in the presence of 1 mM IBMX following incubation with agonists for 30 minutes [2]. Responses to forskolin were examined to serve as an internal control for the adenylyl cyclase assay. [^{3}H]-inositol phosphate production from [^{3}H]-*myo*-inositol-labelled cells was assayed in the presence of 20 mM LiCl following 60 minutes of incubation with agonists [3]. Responses to ATP, acting via an endogenously-expressed purinergic P2Y receptor, were used as an internal control for the inositol phosphate assay. All assays were performed in duplicate.

B. Samuelsson et al. (eds.), Advances in Prostaglandin and Leukotriene Research, 121–123.

Results and Discussion

The IP agonist cicaprost can stimulate both adenylyl cyclase and phospholipase C (PLC) pathways in CHO cells transiently expressing mIP receptors, with pEC_{50} values of 8.35 ± 0.08 and 6.82 ± 0.13 respectively (n ≥ 3). However, although cicaprost (1 μM) produced a similar increase in [^{3}H]-cyclic AMP accumulation in CHO cells transfected with mIP and $IP_{N\text{-}VII}/DP_C$, activation of $IP_{N\text{-}VII}/DP_C$-transfected cells produced merely 30% of the [^{3}H]-inositol phosphate response seen in mIP-transfected cells (Figure 1). It is therefore conceivable that the structural change in $IP_{N\text{-}VII}/DP_C$ that allows improved coupling to G_s, prevents efficient coupling to G_q. In addition, recent studies indicate that the C-terminal tail of the IP receptor is essential for coupling to G_q while having much less effect on the coupling to G_s [4]. Therefore, because the mDP receptor can couple to G_s but not to G_q (Figure 1), then substitution of the C-terminal tail of the mIP receptor with that of the mDP receptor, as in $IP_{N\text{-}VII}/DP_C$, might be expected to produce a receptor less able to couple to G_q.

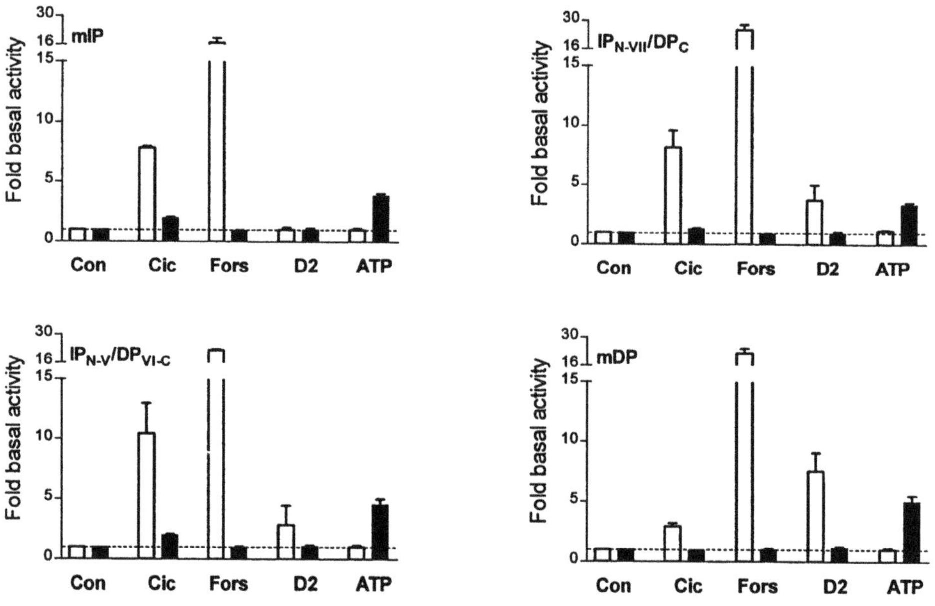

Figure 1. Activation of adenylyl cyclase and phospholipase C pathways by chimeric IP/DP receptors. CHO cells were transiently transfected with mIP (0.5 μg/ml), $IP_{N\text{-}VII}/DP_C$ (0.33 μg/ml), $IP_{N\text{-}V}/DP_{VI\text{-}C}$ (0.4 μg/ml) or mDP (0.5 μg/ml). [^{3}H]-cyclic AMP (open bars) and [^{3}H]-inositol phosphate (filled bars) production was measured in response to stimulation by assay buffer (Con), cicaprost (Cic, 1 μM), forskolin (Fors, 10 μM), PGD_2 (D2, 1 μM) or ATP (100 μM). Results are means ± SEM, n=3.

Unfortunately, our results obtained with the IP_{N-V}/DP_{VI-C}-transfected cells fail to support this conclusion. Surprisingly, cicaprost can stimulate the IP_{N-V}/DP_{VI-C} receptor to increase [^{3}H]-inositol phosphate production to a similar extent as the wild-type mIP receptor (Figure 1). This is despite the lack of the C-terminal tail of the IP receptor (which is reportedly essential for coupling to G_q [4]) and the presence of mDP receptor sequences which do not ordinarily allow coupling to G_q. In addition, a more recent study suggests that the IP receptor is isoprenylated, and removal of the isoprenylation site (i.e. the terminal four amino acid residues in the C-terminal tail) prevents coupling to both adenylyl cyclase and PLC pathways [5]. Clearly our results do not support these latter observations. Therefore, by comparing the ability of mIP, IP_{N-VII}/DP_C, IP_{N-V}/DP_{VI-C} and mDP receptors to couple to adenylyl cyclase and PLC in transiently transfected CHO cells, we can conclude that the C-terminal tail of the mIP receptor is not entirely responsible for its ability to couple to G_q.

Acknowledgments

This work was supported by the Research Grants Council of Hong Kong. The gifts of receptor cDNAs from T. Kobayashi, Department of Pharmacology, Kyoto University, are gratefully acknowledged.

References

1. Wise H, Jones RL. Prostacyclin and its receptors. New York: Kluwer Academic/Plenum Publishers, 2000:1-310.
2. Wise H. Characterization of chimeric prostacyclin/prostaglandin D_2 receptors. Eur J Pharmacol 1999;386:89-96.
3. Conklin BR, Chabre O, Wong YH, Federman AD, Bourne HR. Recombinant $G_q\alpha$. J Biol Chem 1992;267:31-34.
4. Smyth EM, Hong Li W, Fitzgerald GA. Phosphorylation of the prostacyclin receptor during homologous desensitization. J Biol Chem 1998;273:23358-66.
5. Hayes JS, Lawler OA, Walsh M-T, Kinsella BT. The prostacyclin receptor is isoprenylated. J Biol Chem 1999;274:23707-18.

An Unusual Nicotinic Acetylcholine Receptor Activates the 8-Lipoxygenase Pathway in *Aplysia* Neurons

Steven J. Feinmark, Tamara L. Tieman, JacSue Kehoe, and James H. Schwartz

Introduction

Lipoxygenases generate metabolites from arachidonic acid that can act as signaling molecules, both between cells and as second messengers [e.g. 1-3]. In *Aplysia* neurons, 12-lipoxygenase produces substances that mimic the actions of the specific neurotransmitter that stimulated their production. Thus, application of 12-HPETE caused a dual-action response in cell L14 (rapid depolarization followed by slow hyperpolarization), mimicking the responses evoked by histamine [4]. This lipid also produced hyperpolarization in sensory neurons like that induced by the inhibitory neuropeptide FMRFamide [5]. We now show that 8-lipoxygenase products are generated in other *Aplysia* neurons in response to acetylcholine (ACh) [6] and we have characterized the ACh receptor (AChR) pharmacologically.

Results

Application of ACh to *Aplysia* nervous tissue induced the formation of 8-HETE. The effect of ACh was dose-dependent: lipid metabolites were produced at concentrations of the transmitter as low as 1 μM and increased up to 1 mM, the highest dose tested. The response was specific: histamine and FMRFamide (previously shown to activate 12-lipoxygenase in *Aplysia* neurons, [5,7]), 5-HT, GABA, glutamate, octopamine, dopamine, and myomodulin did not induce production of 8-HETE (Figure 1). In addition, depolarization of neurons with 60 mM KCl did not cause the generation of the metabolites (data not shown).

To characterize the receptor that mediates the activation of 8-lipoxygenase, we tested several cholinergic agonists and antagonists previously shown to be effective with *Aplysia* AChRs. Carbachol, a nonspecific cholinergic agonist, was as effective as ACh. Nicotine and suberyldicholine, which activate only the receptors that mediate chloride conductances, also were effective. Of the antagonists tested, only α-BTx blocked activation [8]. Tubocurarine, like atropine, hexamethonium, and TEA failed to inhibit the response. Thus, the AChR that mediates the activation of 8-lipoxygenase appears to be associated with a chloride conductance.

B. Samuelsson et al. (eds.), Advances in Prostaglandin and Leukotriene Research, 125–128.

In *Aplysia*, two pharmacologically distinct receptors have been shown to activate ACh-mediated increases in chloride conductance: one, a rapidly desensitizing chloride conductance that is blocked by the conotoxin, α-CTx-IMI; the other, a sustained chloride conductance that is unaffected by the toxin [9]. α-CTx-IMI failed to block the production of 8-lipoxygenase metabolites induced by ACh or suberyldicholine. Kehoe and McIntosh [9] also showed that the AChR that activates the α-CTx-IMI-insensitive, sustained chloride conductance is differentially distributed in *Aplysia* neurons. In agreement with that, we found that the production of the metabolites occurs only in neurons with receptors that mediate the sustained chloride response. Conversely, identified cells that lack the sustained chloride conductance failed to produce 8-lipoxygenase products.

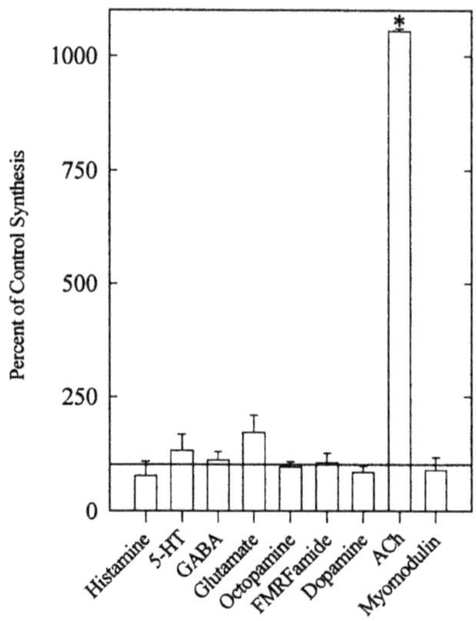

Figure 1. Production of 8-HETE induced by neurotransmitters. *Aplysia* neural components were labeled with [^{3}H]arachidonic acid (10 μCi), washed, and exposed to histamine (50 μM), 5-HT (50 μM), GABA (1 mM), glutamate (50 μM), octopamine (50 μM), FMRFamide (50 μM), dopamine (10 μM), ACh (100 μM), myomodulin (10 μM), or artificial seawater for 10 minutes. These doses are known to elicit physiological responses in *Aplysia*. Lipids were extracted and analyzed by RP-HPLC with a flow-through radioactivity monitor. Baseline release of radioactive products (in the presence of artificial seawater alone) is indicated by the horizontal line. Data are expressed as mean values ± SEM from 3-29 independent experiments; $^*p < 0.05$.

Even though the AChR that mediate the sustained chloride conductance are co-localized with the production of 8-lipoxygenase metabolites, we found that chloride influx alone is not sufficient for activating lipid metabolism. GABA (1 mM) and glutamate (200 μM), both of which induce increases in chloride conductance in *Aplysia* neurons [10], failed to activate 8-lipoxygenase metabolism. We also replaced chloride with an impermeant anion, methanesulfonic acid, in a chloride-free seawater. Under these conditions, ACh failed to elicit influx of chloride but the generation of 8-lipoxygenase products was unaffected (data not shown).

Is a G-protein involved in ACh-induced synthesis of 8-lipoxygenase metabolites? Shapiro et al. [11] showed that the agonist-induced release of arachidonic acid from *Aplysia* neural membrane lipids is blocked by pertussis toxin (PTx). Suberyldicholine-activated production of 8-HETE was significantly reduced in neurons treated with PTx (296 ± 142 cpm versus 53 ± 31cpm; 89% inhibition after correcting for the ASW blank of 22 ± 7 cpm; $p < 0.01$ by repeated measures ANOVA). The release of free arachidonic acid in these experiments also was reduced when the neural components were treated with PTx (39,824 ± 6,840 versus 25,279 ± 3,461; 65% inhibition after correcting for the ASW blank of 17,372 ± 3184 cpm; $p < 0.05$ by repeated measures ANOVA), as expected if the activation of a phospholipase is blocked.

In conclusion, we find that *Aplysia* neurons contain an 8-lipoxygenase that is linked to an ACh-activated chloride channel. Ion flux alone does not result in 8-HETE production. Further, the activation of lipid turnover is inhibited by pertussis toxin suggesting that a G-protein plays a role in regulating ACh-dependent arachidonate metabolism in these neurons.

Acknowledgments

This work was supported by NS29832, MH00921, CNRS UMR 8544.

References

1. Samuelsson B. Leukotrienes: mediators of immediate hypersensitivity reactions and inflammation. Science 1983;220:568-75.
2. Shimizu T, Wolfe LS. Arachidonic acid cascade and signal transduction. J Neurochem 1990;55:1-15.
3. Feinmark SJ, Piomelli D, Shapiro E, Schwartz JH. Biologically active metabolites of the 12-lipoxygenase pathway are formed by *Aplysia* nervous tissue. Ann NY Acad Sci 1989; 559:121-30.
4. Piomelli D, Shapiro E, Zipkin R, Schwartz JH, Feinmark SJ. Formation and action of 8-hydroxy-11,12-epoxy-5,9,14-icosatrienoic acid in *Aplysia*: A possible second messenger in neurons. Proc Natl Acad Sci USA 1989;86:1721-25.
5. Piomelli D, Volterra A, Dale N, et al. Lipoxygenase metabolites of arachidonic acid as second messengers for presynaptic inhibition of *Aplysia* sensory cells. Nature 1987;328: 38-43.
6. Steel DJ, Tieman TL, Schwartz JH, Feinmark SJ. Identification of an 8-lipoxygenase pathway in nervous tissue of *Aplysia californica*. J Biol Chem 1997;272:18673-81.
7. Piomelli D, Shapiro E, Feinmark SJ, Schwartz JH. Metabolites of arachidonic acid in the nervous system of *Aplysia*: possible mediators of synaptic modulation. J Neurosci 1987;7:

3675-86.

8. Kehoe J, Sealock R, Bon C. Effects of α-toxins from *Bungarus multicinctus* and *Bungarus caeruleus* on cholinergic responses in *Aplysia* neurones. Brain Res 1976;107:527-40.
9. Kehoe J, McIntosh JM. Two distinct nicotinic receptors, one pharmacologically similar to the vertebrate α7-containing receptor, mediate Cl currents in *Aplysia* neurons. J Neurosci 1998;18:8198-213.
10. King WM, Carpenter DO. Voltage-clamp characterization of Cl^- conductance gated by GABA and L-glutamate in single neurons of *Aplysia*. J Neurophysiol 1989;61:892-99.
11. Shapiro E, Piomelli D, Feinmark SJ, Vogel S, Chin G, Schwartz JH. The role of arachidonic acid metabolites in signal transduction in an identified neural network mediating presynaptic inhibition in *Aplysia*. Cold Spring Harbor Symp Quant Biol 1988; 53:425-33.

Neutrophil-Endothelial Cell Interaction: A Multidirectional Cross-talk in Leukotriene Generation

Giancarlo Folco

Introduction

The interaction between neutrophils (PMNLs)-monocytes and endothelial cells (ECs) represents the initial event in the acute inflammatory response and in the pathogenesis of vascular diseases [1]. The vascular endothelium represents a dynamically mutable interface that may undergo phenotypic modulations to a pathophysiologically dysfunctional state, characterized by a loss of critical homeostatic mechanisms present in healthy cells (e.g. reduced NO and PGI_2 synthesis, increased expression of adhesive molecules ligands) [2]. On the other hand, it has been repeatedly demonstrated that PMNLs activation, increased adhesion to ECs and tissue accumulation aggravate ischemic tissue damage in brain and heart and leukocyte depletion may decrease tissue necrosis [3]. Both PMNLs and ECs generate an impressive repertoire of biological effectors among which lipid mediators (e.g. eicosanoids: prostaglandins (PGs) and leukotrienes (cys-LT)) have attracted considerable interest. Their generation exhibits remarkable cellular specificity; however, their formation may also occur through transfer of reactive intermediates between adjacent cells (i.e. transcellular biosynthesis) which represents a specialized mode of cell communication [4]. This process suggests that the cellular environment (i.e. cell-cell interactions) represents a fundamental control mechanism in the production of lipids that may ultimately affect vascular function [5].

The cooperative formation of cys-LT has been shown to cause coronary vasoconstriction and severe inflammatory changes in the rabbit heart [6-7]. However, previous studies have suggested that the transcellular biosynthesis process is even more complicated and may often involve an additional exogenous supply of arachidonate from the platelet or endothelial cell, which is taken up by the neutrophil for conversion to the LTA_4 that is then handed back to the donor cell for final conversion to cys-LT [8].

In order to gain further insight into the biochemistry of this neutrophil-endothelial cell interaction for generation of bioactive cys-LT, we investigated the metabolic profile of LTA_4-derived metabolites in isolated neutrophils as well as when perfused through a spontaneously beating, isolated rabbit heart. The study demonstrates that the metabolic fate of arachidonate released from the endogenous pool of neutrophil phospholipids is different from that of arachidonate derived from exogenous sources (e.g. endothelial cells).

B. Samuelsson et al. (eds.), Advances in Prostaglandin and Leukotriene Research, 129–133.

Methods

An isolated, coronary-perfused whole heart preparation from rabbit has been used [6], in the presence or absence of circulating cells (e.g PMNLs), at constant flow (20ml/min, respectively), so that changes in vascular resistance can translate into increases or decreases of perfusion pressure as monitored by a pressure transducer. The perfusing fluid has been collected for HPLC or enzyme-immunoassay (EIA)of lipid mediator formation, which has been compared with that obtained in PMNLs or vascular cells alone [9].

Results and Discussion

fMLP (0.3 μM) challenge of neutrophils (10^5 cells/ml) in the reperfused, spontaneously beating, isolated rabbit heart, resulted in the production of large amounts of cys-LT, presumably as a result of neutrophil-endothelial cell cooperative synthesis of cys-LT. Quantitative analysis revealed that in the neutrophil-perfused hearts the amount of LTA_4 metabolites (enzymatic+nonenzymatic) was much more than that seen when the same number of neutrophils were stimulated *in vitro* (Figure 1). The data suggested that the interaction of neutrophils with the coronary vasculature of the heart not only led to the new synthesis of cys-LTs (transcellular biosynthesis) but also to a fourfold increase in total LTA_4 metabolites, likely because of additional sources of externally derived (non-neutrophil) arachidonate.

One possible explanation for the preferential release of LTA_4, (available for cys-LT formation), rather than conversion into LTB_4, was that the LTA_4-hydrolase might have been saturated with substrate. The relative ratio of enzymatic to nonenzymatic metabolites of LTA_4 resulting from the addition of increasing amounts of arachidonate in unprimed neutrophils was investigated and found to remain unchanged even when the concentration of exogenous arachidonate varied (1-10 μM).

Moreover, when unprimed neutrophils were challenged with fMLP and exogenous arachidonate (2-5 μM) more nonenzymatic LTA_4 metabolites were produced than seen with GM-CSF primed cells stimulated with fMLP in the absence of added arachidonate, and the ratio of nonenzymatic to enzymatic products showed a twofold increase. In order to determine if the increased production of nonenzymatic metabolites reflected an increased availability of intact LTA_4 for transcellular metabolism, GM-CSF primed neutrophils and platelets were co-incubated at a ratio of 1:40 (neutrophil:platelet) and challenged with fMLP (which does not activate the platelets) in the presence of deuterium-labeled arachidonate (5 μM). This resulted in the preferential formation of labeled LTC_4, when compared to enzymatic metabolites of LTA_4-hydrolase.

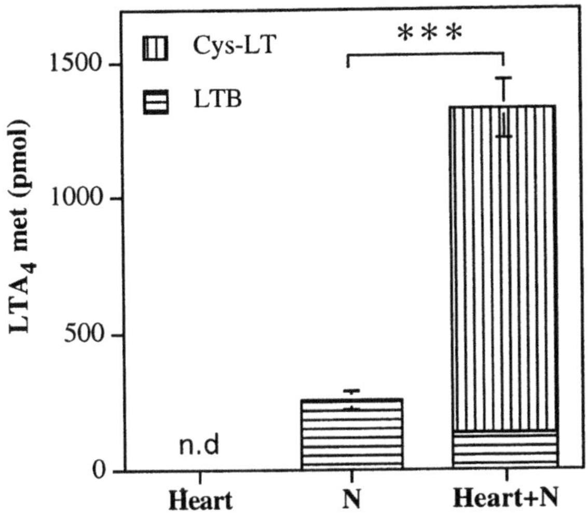

Figure 1. Increased production of LTA_4 metabolites by combination of GM-CSF primed human neutrophils and isolated rabbit hearts. Effect of challenge with fMLP (0.3 μM) on the production of LTA_4 metabolites by isolated rabbit hearts, by neutrophils in suspension, and by neutrophils perfusing the isolated rabbit heart under recirculation conditions. Purified human neutrophils (10^7 cells) were primed with GM-CSF (1 nM, 30 minutes) prior to perfusion. LTA_4 metabolites were measured by RP-HPLC. Values are expressed as mean ± S.E. (*n* =3-5). ***, $p < 0.001$. N, neutrophils. From J Biol Chem 1999;274:28264-69.

Transcellular biosynthesis of cys-LT involving cooperation between circulating and vascular cells, represents a critical mechanism in the synthesis of leukotrienes that may determine the overall quantity, as well as profile, of arachidonate metabolites generated by cell activation at the organ level [10-11].

fMLP stimulation of GM-CSF primed human neutrophils, circulating within the coronary vascular bed, resulted in the production of very large amounts of cys-LT; in sharp contrast, the same numbers of GM-CSF primed neutrophils stimulated with fMLP *in vitro* produced significantly lower amounts of overall LTA_4 metabolites and only LTA_4 hydrolase-derived products, namely LTB_4 and its ω-oxidized derivatives, while LTC_4 as well as nonenzymatic LTA_4 metabolites could not be observed. This apparent discrepancy could imply that additional arachidonate was provided to the neutrophils by rabbit heart cells, likely endothelial cells, thereby significantly enhancing the production of leukotrienes. Furthermore it appeared likely that this additional arachidonate supplied to neutrophils was undergoing a preferential metabolism into a pool of LTA_4 that was readily exported outside of the neutrophil and therefore available for transcellular metabolism into LTC_4 by adjacent endothelial cells (Figure 2).

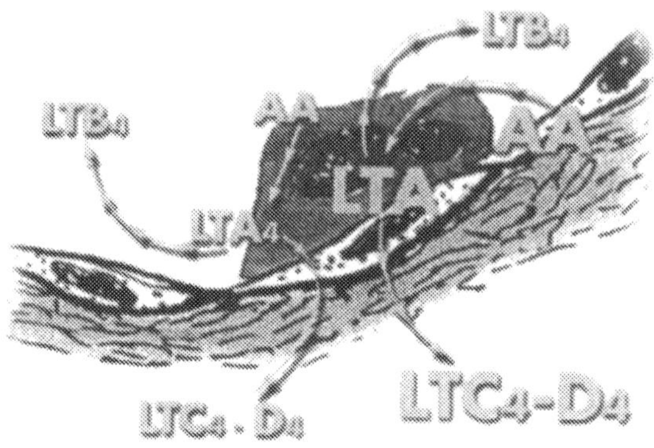

Figure 2. Neutrophil-endothelial cell interactions: a multidirectional cross-talk in eicosanoid generation.

Indeed, recent results [12] showed that significantly larger amounts of the exogenous arachidonate underwent preferential metabolism to an LTA_4 pool that was not apparently available to cytosolic LTA_4 hydrolase, thus resulting in export and nonenzymatic hydrolysis. Moreover, the unstable epoxide exported from the donor PMNLs, was shown to be of biological relevance, since, in the presence of a limited number of platelets, the LTA_4 arising from metabolism of exogenous arachidonate was preferentially converted into LTC_4 by the acceptor platelets (rather than into LTB_4) as a result of transcellular cooperation between neutrophils and platelets [12].

These results may suggest that exogenous arachidonate is metabolized to leukotrienes by a pathway that is physically separate in the cell from that involved in conversion of arachidonate from endogenous sources, most likely by a portion of the 5-lipoxygenase located at (or translocated to) the plasma membrane where the enzyme preferentially metabolizes arachidonate reaching the cell from the outside.LTA_4 produced at this site would be available for export and may have some difficulty gaining access to the cytosolicLTA_4 hydrolase.

Whatever the mechanism, these observations suggest that transcellular biosynthesis of leukotrienes is characterized by different interesting features: a) it represents a unique pathway of arachidonate metabolism and eicosanoid generation; b) it occurs with physiologic cell stimuli; c) it can be extended to an intact organ such as a spontaneously beating, isolated heart; and d) it may involve complex multidirectional cross-talks between cells – namely arachidonate from endothelial cells to neutrophils and LTA_4 back to the endothelial cells, with significant physiopathologic consequences.

References

1. Malech HL, Gallin JI. Neutrophils in human disease. N Engl J Med 1987;317:687.
2. Lefer AM, Lefer DJ. Pharmacology of the endothelium in ischemia-reperfusion and circulatory shock. Ann Rev Pharmacol Toxicol 1993;33:71-90.
3. Lucchesi BR. Modulation of leukocyte-mediated myocardial reperfusion. Ann Rev Physiol 1990;52:561-76.
4. Maclouf J, Murphy RC, Henson P. Transcellular sulfidopeptide leukotriene biosynthetic capacity of vascular cells. Blood 1989;74:703-7.
5. Maclouf J, Folco GC, Patrono C. Eicosanoids and isoeicosanoids: constitutive, inducible and transcellular biosynthesis in vascular disease. Thromb Haemost 1998;79:691-705.
6. Sala A, Rossoni G, Buccellati C, Berti F, Folco GC, Maclouf J. Formation of sulfidopeptide-leukotrienes by cell-cell interaction causes coronary vasoconstriction in isolated, cell-perfused heart of rabbit. Brit J Pharmacol 1993;110:1206-12.
7. Sala A, Aliev G, Rossoni G, et al. Morphological and functional changes of coronary vasculature caused by transcellular biosynthesis of sulfidopeptide leukotrienes in isolated heart of rabbit. Blood 1996;87:1824-32.
8. Antoine C, Murphy RC, Henson PM, Maclouf J. Time-dependent utilization of platelet arachidonic acid by the neutrophils in formation of 5-lipoxygnase products in platelet-neutrophil co-incubations. Biochim Biophys Acta 1992;1128:139-46
9. Sala A, Bolla M, Zarini S, Muller-Peddinghaus R, Folco GC. Release of leukotriene A_4 versus leukotriene B_4 from human polymorphonuclear leukocytes. J Biol Chem1996;271:17944-48.
10. Brady HR, Serhan CN. Adhesion promotes transcellular leukotriene biosynthesis during neutrophil-glomerular endothelial cell interactions: inhibition by antibodies against CD18 and L-selectin. Biochem Biophys Res Commun 1992;186:1307-14.
11. Rossoni G, Sala A, Berti F, et al. Myocardial protection by the leukotriene synthesis inhibitor BAY X1005; importance of transcellular biosynthesis of cysteinyl leukotrienes. J Pharmacol Exp Therap 1996;276:335-41.
12. Sala A, Zarini S, Folco GC, Murphy RC, Henson P. Differential metabolism of exogenous and endogenous arachidonic acid in human neutrophils. J Biol Chem 1999;274(40):28264-69.

Cyclooxygenase in the Lamb Ductus Arteriosus: Developmental Changes and Upregulation

Flavio Coceani, Cameron Ackerley, Eric Seidlitz, Lois Kelsey

Introduction

The ductus arteriosus is kept patent in the fetus by a process involving prostaglandin (PG) E_2, both local and blood-borne, as the main effector [1]. The PGE_2-based relaxing mechanism develops early in gestation and, in fact, with a premature birth its intramural component may prevail over the oxygen constriction [1]. This knowledge has, on one hand, introduced indomethacin in the management of prematurely born infants with persistent ductus and, on the other hand, has highlighted potential detrimental effects on the unborn child of nonsteroidal anti-inflammatory drugs taken by the mother. The realization now that PGE_2 may originate from two cyclooxygenase (COX) isoforms, COX1 and COX2, has added complexity but, at the same time, has broadened the scope of any therapeutic intervention.

Our aim was to gain a better insight into the functional organization of the COX system in the ductus through the last third of gestation. We looked for the two COX isozymes in the tissue and determined concomitantly the response of the vessel to a selective COX2 inhibitor (L-745,337) [2] versus a nonselective COX1/COX2 inhibitor (indomethacin). Experiments were performed in naive preparations and preparations that had been treated with agents, such as oxygen and endotoxin, which may promote COX function and are also relevant to the clinical situation.

Methods

Preterm (103-107 days, 0.7 gestation; 94-97 days, 0.65 gestation) and near-term (134-139 days gestation; term, 145 days) pregnant sheep were used. Procedures for anesthesia, Cesarean delivery of fetuses, and isolation of the ductus arteriosus have been described [3]. However, in this study certain fetuses, both term and preterm (0.7 gestation), were given endotoxin *in utero* (0.1 $\mu g\ kg^{-1}$ IV) three hours prior to delivery.

When analyzed morphologically, ductus specimens were processed immediately or after treatment with endotoxin *in vitro* (100 ng ml^{-1}, 2 hours). For immunofluorescence microscopy, they were fixed, and sections were incubated first with an antibody against COX1 or COX2 (dilution, 1:200) and then with an appropriate secondary antibody. For

B. Samuelsson et al. (eds.), Advances in Prostaglandin and Leukotriene Research, 135–137.

transmission immunoelectron microscopy, ultrathin sections were incubated with the primary antibody and a gold-labeled, secondary antibody (particle, 10 nm).

For the pharmacological study, preparations were mounted in an organ bath and gassed with 2.5% O_2:5% CO_2 in N_2 to mimic the condition *in utero* [3]. The neonatal condition was reproduced by bubbling increasing oxygen concentrations (15%, 30%, 95%). L-745,337 was tested cumulatively (1, 2.8, 10 μM) and was compared to indomethacin. Experiments were carried out in untreated and endotoxin-treated (*in vitro* and *in vivo*) preparations.

Results

Epifluorescence microscopy showed COX1 immunolabeling in endothelial and smooth muscle cells. No major difference was noted between term and preterm (0.7 gestation), nor was any obvious increase observed with endotoxin *in vivo*. Conversely, COX2 immunoreactivity was barely detectable in the premature and remained relatively weak at term. However, contrary to COX1, COX2 labeling increased after giving endotoxin *in vivo* at either age. On transmission electron microscopy, COX1-linked gold particles exceeded COX2-linked particles at all ages. In addition, both labels increased with gestation. There was also no change in total COX1 immunogold reactivity, or at most a marginal increase, in endotoxin-treated tissues (*in vitro* and *in vivo*). COX2, however, responded differently to endotoxin depending on whether treatment was *in vitro* or *in vivo*. In the former case, there was a peculiar loss of the enzyme, while in the latter case there was the expected upregulation, particularly at term. The reason for such loss is speculative. By acting protractedly on the tissue, endotoxin might have caused excess NO formation and, hence, downregulation of COX2 after an initial upregulation [4]. Alternatively, or concomitantly, downregulation may have resulted from activation of the PGE_2 receptor [5]. Whether stimulatory or not, endotoxin brought a redistribution of both COX1 and COX2 from the inner to the outer regions of the cytoplasm.

L-745,337 contracted the term ductus dose-dependently and the response nearly equaled that to indomethacin. By contrast, at 0.7 gestation L-745,337 was less effective (60% reduction) than indomethacin. In the youngest fetuses (0.65 gestation) the contraction to the COX2 inhibitor was barely visible. Premature preparations treated with endotoxin *in vitro* tended to contract less to L-745,337, while they contracted more after treatment *in vivo*. The same preparations responded more to the inhibitor after being primed with oxygen. The oxygen contraction, on the other hand, was increased in L-745,337-treated tissues. Hence, we conclude that oxygen promotes COX2 in the premature and that COX2, via PGE_2, curtails the oxygen contraction.

Conclusion

Our study demonstrates that COX1 and COX2 occur constitutively in the ductus. Their development is uneven through gestation, with COX2 contributing little to PGE_2 formation in the premature and matching instead COX1 at term. However, before term COX2 may acquire greater importance under the influence of oxygen and endotoxin. When considered in a clinical context, these data reaffirm the usefulness of a dual COX1/COX2 inhibitor for a persistent ductus in the premature infant. Supporting this conclusion is the expected exposure of the ductus to conditions (postnatal rise in blood oxygen tension, mechanical stress, any intervening infection) resulting in COX2 upregulation. Conversely, a COX2 inhibitor appears a better choice for preventing premature labor, since it may combine the desired effect with greater safety for the fetus. Specifically, by affecting little or not at all the ductus, the pulmonary vasculature, and blood levels of PGE_2, such inhibitor can ensure a smoother transition from intra- to extrauterine life.

Acknowledgment

Supported by the Heart and Stroke Foundation of Ontario (Grant T-3329).

References

1. Smith GCS. The pharmacology of the ductus arteriosus. Pharmacol Rev 1998;50:35-58.
2. Chang C-C, Boyce S, Brideau C, et al. Pharmacology of a selective cyclooxygenase-2 inhibitor, L-745, 337: a novel anti-inflammatory agent with an ulcerogenic sparing effect in rat and nonhuman primate stomach. J Pharmacol Exp Ther 1995;274:1531-37.
3. Coceani F, Huhtanen D, Hamilton NC, Bishai I, Olley PM. Involvement of intramural prostaglandin E_2 in prenatal patency of the lamb ductus arteriosus. Can J Physiol Pharmacol 1986;64:737-44.
4. Swierkosz TA, Mitchell JA, Warner TD, Botting RM, Vane JR. Co-induction of nitric oxide synthase and cyclooxygenase: interactions between nitric oxide and prostanoids. Br J Pharmacol 1995;114:1335-42.
5. Bishop-Bailey D, Pepper JR, Larkin SW, Mitchell JA. Differential induction of cyclooxygenase-2 in human arterial and venous smooth muscle - Role of endogenous prostanoids. Arterioscler Thromb Vasc Biol 1998;18:1655-61.

Co-localization of Prostacyclin (PGI2)-Synthase and Caveolin-1 in Endothelial Cells Underscores New Roles of PGI2 in Angiogenesis

Enzo Spisni, Cristiana Griffoni, Spartaco Santi, Massimo Riccio, Roberta Marulli and Vittorio Tomasi

Introduction

In vascular cells, PGI2 synthase (PGI2s) has been localized in the endoplasmic reticulum of endothelial cells and in the nuclear and plasma membrane of smooth muscle cells (SMC). In human umbilical vein endothelial (HUVE) cells, a direct confocal microscopy approach indicates that more than 80% of the enzyme resides in cellular sites co-staining with caveolin-1 antibody. This evidence was confirmed by the demonstration that PGI2 synthase and caveolin-1 are associated to detergent-insoluble membrane domains in the same low density fractions of a sucrose gradient and the depletion of cellular cholesterol leads to the shift of PGI2s and caveolin-1 to higher density fractions of the gradient. Moreover, a double approach, based on the usage of filipin as a specific caveolae disrupting agent and antisense oligonucleotides (ODNs) targeting PGI2 synthase mRNA, suggests that the production of PGI2 in caveolae is likely to be connected to the regulation of angiogenesis, at least *in vitro*.

Materials and Methods

Sucrose Gradient

Protein separation on sucrose gradient was performed following the modifications proposed by Zurzolo et al. [1] to the method first described by Brown and Rose [2]. Briefly, four confluent 75 cm^2 flasks of HUVE cells were rinsed with PBS and lysed in 1 ml of TNE/TX100 buffer (25 mM Tris-Cl [pH 7.5], 150 mM NaCl, 5 mM EDTA, 1% Triton X-100). A linear sucrose gradient (5-35% in TNE buffer) was layered over the lysate and centrifuged for 18 hours at 200,000 x g at 4°C. Fractions were collected and TCA-precipitated proteins were analyzed by Western blot to detect cav-1 and PGI2s.

Cholesterol Depletion

Depletion of cholesterol was performed following the method described by Keller and

B. Samuelsson et al. (eds.), Advances in Prostaglandin and Leukotriene Research, 139–142.

Simons [3].

SYNTHESIS OF ODNS

Phosphorothioate ODNs were synthesized using an Applied Biosystem 394 DNA/RNA Synthesizer, according to manufacturer's instructions. They had the following sequences: anti-PGI2 synthase exon 1 5'-AGTAGCAGCAGCAGCAA-3'; anti-PGI2 synthase 3'UTR 5'-CTTATCTGAATAGCATTTGTGG-3'; sense oligo, 5'-TTGCTGCTGCTGCTACT-3'; scrambled oligo 5'-AGACATCGAAGCACGAG-3'.

3-D ANGIOGENESIS IN FIBRIN GEL

To evaluate the effect of Filipin or anti-PGI2 synthase ODNs on *in vitro* angiogenesis, HUVEC were seeded on 24 well plates coated with fibrin gel, prepared as described by Montesano et al. [4]. Filipin (10-70 ng/ml) or ODNs (1 μM) were added directly into the medium. Cells were observed at the light microscope 24 hours later.

Results and Discussion

We have previously immunodetected PGI2 synthase in HUVE cells on abundant cytoplasmic vesicles whose distribution resembled that of vesicles originating from surface caveolae [5]. Gold-albumin, that binds to a caveolar receptor [6], revealed a vesicular distribution in HUVEC cytoplasm similar to that observed during PGI2 synthase immunodetection and the confocal microscopical analysis of caveolin-1 and PGI2s distribution confirmed that the two proteins were strongly co-localized (data not shown). Moreover, we detected PGI2s and caveolin-1 at the same low density fractions of a 5%-35% sucrose gradient (Figure 1, panel A1) and we observed a missorting of PGI2s and caveolin-1 in HUVEC treated with mevalonate and methyl-β-cyclodextrin (Figure 1, panel A2). This treatment reduces cellular cholesterol levels by 60-70%, leading to the disruption of caveolae and missorting of rafts-associated proteins [3]. The shift of both caveolin-1 and PGI2s towards the same high density fractions of the sucrose gradient, following cholesterol depletion, supports their association.

By using filipin, a caveolae disrupting agent, we demonstrated that prostacyclin production in IL-1α stimulated HUVEC was significantly decreased (not shown) and capillary-like tube formation in fibrin gel was impaired (Figure 1, panels B1-B3). Moreover, the treatment of HUVEC with two antisense ODNs directed against PGI2s mRNA at the same time affected PGI2s levels and enzymatic activity (not shown) and inhibited capillary-like tube formation in fibrin gel (Figure 1, panels C1-C3).

This observation underlines a role of PGI2s as angiogenic enzyme and suggests the involvement of caveolae in angiogenesis. The association of PGI2 synthase with caveolin-1 could allow the shuttling of this enzyme towards the perinuclear space, where cycloxygenase is present [7], targeting PGI2 production in the proximity of its nuclear receptor belonging to peroxisomal proliferator-activated receptors (PPAR-δ). The activation of PPAR-δ by a PGI2 stable agonist has been observed during embryo implantation in the mouse, an event requiring angiogenesis [8].

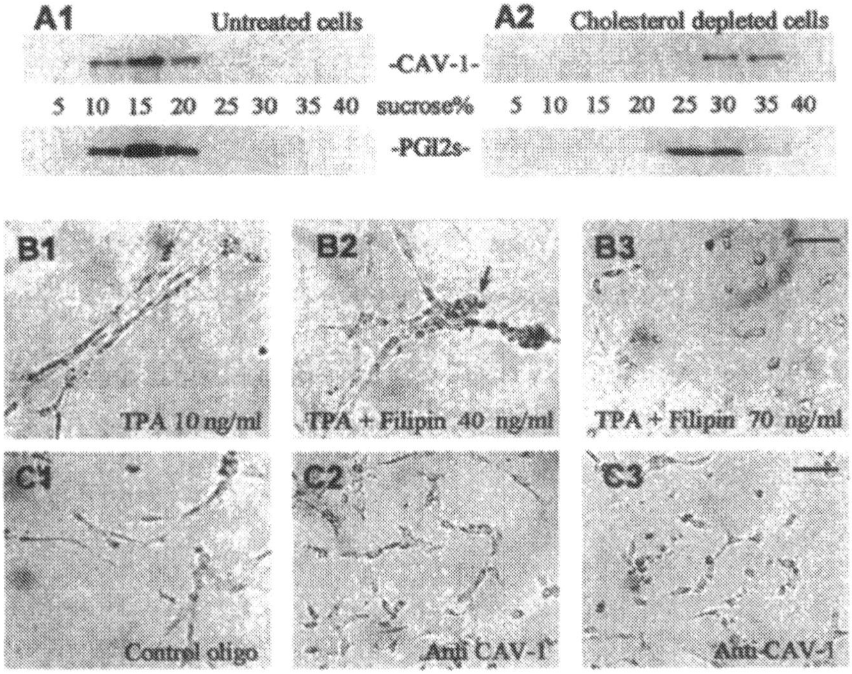

Figure 1. Panels A1-A2: fractions collected from the top to the bottom of a sucrose gradient were analyzed by Western blot. Fluorograms represent the distribution of PGI2s and cav-1 in untreated HUVEC (panel A1) and in HUVEC treated with 10 μM mevalonate and 10 μM methyl-β-cyclodextrin to obtain cholesterol depletion (panel A2). Panels B1-B3: effect of filipin on 3D angiogenesis test in the presence of TPA (10 ng/ml). Panel B1 shows the control after 24 hours of incubation; panel B2 shows the effect of 40 ng/ml filipin, arrow points to clusters of cells which are not competent for capillary morphogenesis; panel B3 shows that 70 ng/ml filipin completely inhibits tube formation. Bar = 100 μM. Panels C1-C4: effect of anti-PGI2 synthase ODNs on 3D angiogenesis test. C1, HUVEC treated with PGI2s sense oligo; C2, HUVEC treated with anti PGI2s 3'-UTR oligo; C3, HUVEC treated with anti PGI2s exon 1 oligo. Bar = 100 μM.

References

1. Zurzolo C, van't Hof W, van Meer G, Rodriguez-Boulan E. VIP 21/caveolin, glycosphingolipid clusters and sorting of glycophosphatilinositol-anchored proteins in

epithelial cells. EMBO J 1994;13:42-53.

2. Brown DA, Rose JK. Sorting of GPI-anchored proteins to glycolipid-enriched membrane subdomains during transport to the apical cell surface. Cell 1992;68:533-44.
3. Keller P, Simons K. Cholesterol is required for surface transport of influenza virus hemagglutinin. J Cell Biol 1998;140:1357-67.
4. Montesano R, Pepper M S, Mohle-Steinlein U, Risau W, Wagner EF, Orci L. Increased proteolytic activity is responsible for the aberrant morphogenetic behavior of endothelial cells expressing the middle T oncogene. Cell 1990;62:435-45.
5. Spisni E, Bartolini G, Orlandi M, Belletti B, Santi S, Tomasi V. Prostacyclin (PGI2) synthase is a constitutively expressed enzyme in human endothelial cells. Exp Cell Res 1995;219:507-13.
6. Tiruppathi C, Finnegan A, Malik AB. Isolation and characterization of a cell surface albumin-binding protein from vascular endothelial cells. Proc Natl Acad Sci USA 1996; 93:250-54.
7. Spencer AG, Woods JW, Arakawa T, Singer II, Smth WL.. Subcellular localization of prostaglandin endoperoxide synthase-1 and –2 by immunoelectron microscopy. J Biol Chem 1998;273:9886-93.
8. Lim H, Gupta RA, Ma W, et al. Cyclo-oxygenase-2-derived prostacyclin mediates embryo implantation in the mouse via PPAR d. Genes Dev 1999;13:1561-74.

LEUKOTRIENE C_4 SYNTHASE GENE POLYMORPHISM IN SEVERE ASTHMA

Anthony Sampson, Ian Sayers, Peter Howarth, and Stephen Holgate

Introduction

The cysteinyl-leukotrienes (cys-LT) LTC_4, LTD_4, and LTE_4 are well-established as mediators of bronchoconstriction and eosinophilia in asthma [1-3]. We have shown that in the asthmatic bronchial wall, eosinophils also represent the majority of cells that express the enzymes of the cys-LT pathway and hence have the capacity to generate cys-LT upon stimulation [4,5]. Inflammation and bronchoconstriction in the asthmatic airway may therefore be sustained by a vicious cycle of cys-LT synthesis and eosinophil recruitment. Anomalies in the regulation of the cys-LT pathway in eosinophils may thus be central to the pathophysiology of asthma.

The capacity of eosinophils and other leukocytes to synthesize cys-LT may be influenced by promoter polymorphism in the genes encoding cys-LT pathway enzymes [6]. Cys-LT are synthesized from membrane-derived arachidonic acid by the sequential actions of 5-lipoxygenase (5-LO), 5-LO activating protein (FLAP), and LTC_4 synthase. Promoter polymorphism in 5-LO is linked to reduced 5-LO gene transcription and a diminished clinical response to an oral LT synthesis inhibitor in a small proportion (~5%) of asthmatics [7,8], but this cannot account for the large sub-group (30-40%) of relatively poor responders to anti-LT therapy.

In bronchial biopsies of patients with aspirin-intolerant asthma (AIA), who have constitutively high baseline production of cys-LT and generally good clinical responses to leukotriene-modifier drugs [9], we have shown that the numbers of cells expressing LTC_4 synthase are 5-fold higher than in biopsies from aspirin-tolerant asthmatics (ATA) and 19-fold more prevalent than in normal bronchial biopsies [4]. The increased prevalence of LTC_4 synthase in AIA biopsies was due partly to a 2-fold bronchial eosinophilia and partly to an increased proportion of eosinophils expressing the enzyme (50% versus 21% in ATA patients). LTC_4 synthase over-expression correlated exclusively with elevated cys-LT levels in BAL fluid and with bronchial hyperresponsiveness to inhaled lysine-aspirin [4]. A single nucleotide polymorphism (SNP) in the LTC_4 synthase gene promoter ($A_{-444}C$) has been reported in 76% of a group of Polish AIA patients compared to 42-44% of ATA and normal subjects [10]. The variant allele (C_{-444}) was associated with higher levels of LTC_4 synthase mRNA in blood eosinophils [11], suggesting that the mutant allele increases expression of LTC_4 synthase protein and hence enhances the capacity of these cells to synthesise cys-LT.

B. Samuelsson et al. (eds.), Advances in Prostaglandin and Leukotriene Research, 143–148.

However, other workers found no difference in the prevalence of the LTC_4 synthase C_{-444} SNP in American AIA patients [12].

Since the variant LTC_4 synthase genotypes (C/C and C/A) are also found in a significant proportion (42-44%) of ATA patients and normal subjects [10], we reasoned that the variant LTC_4 synthase allele may not be associated specifically with AIA, but rather with an increased contribution of cys-LT to asthma of any clinical phenotype. We hypothesized firstly that eosinophils from subjects with the variant genotypes may have a predisposition to enhanced synthesis of cys-LT when the 5-LO pathway is stimulated. Secondly, we hypothesized that the variant LTC_4 synthase genotypes may be more prevalent in severe persistent asthmatics, irrespective of aspirin sensitivity, than in normal subjects. Thirdly, we hypothesized that the variant genotypes may identify a sub-group of severe asthmatic patients in whom elevated cys-LT synthesis is particularly important in impairing lung function, and hence that these patients may be synonymous with a clinical phenotype of good responders to oral anti-LT therapy with zafirlukast. Our results show that the variant LTC_4 synthase genotypes are linked to elevated *ex vivo* synthesis of LTC_4 by blood eosinophils, are more prevalent in severe chronic asthmatics than in normal subjects, and tend to predict lung function response to oral therapy with a LT receptor antagonist.

Methods

We assessed urinary LTE_4 and *ex vivo* LTC_4 synthesis by immunomagnetically-purified blood eosinophils in 31 normal subjects, and lung function response to oral zafirlukast in 23 severe chronic asthmatics. All subjects were genotyped for the $A_{-444}C$ polymorphism in LTC_4 synthase. Normal subjects (n=31) were adult volunteers with no history of chronic respiratory disease and taking no medication. Four were atopic as judged by skin-prick tests. The 23 asthmatic patients (6M,17F; aged 25-76 yr) had a clinical history of poorly-controlled chronic severe asthma. Three asthmatics had a clinical history of aspirin sensitivity. All asthmatic patients were receiving inhaled corticosteroids, and 10 were also receiving oral steroids, 14 a long-acting β_2-agonist, four theophylline, and one cyclosporin. At pre-study assessment, FEV_1 varied between 0.24 and 2.5 l and PEF between 106 and 408 l/min. For one week before and throughout the study, asthmatics abstained from all medications except inhaled corticosteroids and inhaled β_2-agonists prn, and received oral zafirlukast (20 mg bd) for two weeks, when lung function assessments were repeated. Although the trial was an open study, patients and investigators were blinded to patient genotype. The study was approved by the Southampton & SouthWest Hampshire Joint Research Ethics Committee, and all subjects gave informed written consent.

Eosinophils were purified immunomagnetically from the blood of normal subjects as described [13]. Eosinophils (> 98% pure, > 96% viable) were cultured (10^5 cells/0.5 ml) in HEPES-buffered medium at 37°C for 15 minutes in the presence and absence of calcium ionophore A23187 (1 μM) and in the presence of indomethacin (10 μM). Total cys-LT synthesis was quantified by Biotrak enzyme immunoassay (EIA) (Amersham). Urinary

LTE_4 levels were assessed by EIA as a marker of whole-body cys-LT synthesis after correction for creatinine. Genomic DNA was isolated from whole blood using the QIAmp DNA Blood Mini Kit (Qiagen). The region flanking the $A_{-444}C$ polymorphism was amplified using PCR (forward primer: 5' CTC CAT TCT GAA GCC AAA G 3' ; reverse primer: 5' AGA CCG CCT CAC CAC TT 3'). Alleles were genotyped using MspI restriction digestion of the 296bp PCR product [10]; an endogenous MspI site provided an internal control for the assay.

Results

The variant LTC_4 synthase genotypes (C/C and C/A) were observed in 13 out of 23 severe asthmatic patients (56%) compared to 10 out of 31 normal subjects (32%), giving frequencies for the variant allele (C_{-444}) of 0.27 in the asthmatics and 0.19 in the normals. Despite the small size of the study group, the enhanced prevalence of the variant genotype in the asthmatics was statistically significant (P=0.04, Fisher's exact test).

In normal subjects, spontaneous production of LTC_4 from unstimulated blood eosinophils was not significantly different between the two genotype sub-groups. However, when stimulated with calcium ionophore A23187 (1 uM) in the presence of indomethacin (10 uM) to block shunting of arachidonate substrate along the prostanoid pathway, eosinophils from patients with variant genotypes (C/C and C/A) produced 3-fold more LTC_4 (7.8 ± 2.9 ng/10^6 cells, n=13) than eosinophils from patients with the wild-type genotype (A/A, 2.7 ± 0.4 ng/10^6 cells, n=10; P=0.04 Mann-Whitney) (Figure 1). There was no significant difference between normal subjects with the variant genotypes and those with the wild-type genotype in urinary levels of LTE_4 (145 ± 21 versus 128 ± 8 pmol/mmol.Creat. respectively; $P > 0.2$) (Figure 1).

In asthmatic subjects with variant LTC_4 synthase genotypes, oral zafirlukast (20 mg bd) for two weeks increased FEV_1 by $9 \pm 12\%$ and FVC by $15 \pm 18\%$ compared to pre-study values (n=13), with decreases of $-12 \pm 18\%$ and $-18 \pm 15\%$ respectively in patients with the wild-type genotype (n=10; both P=0.1) (Figure 2). Home PEF values did not differ between the groups.

Discussion

The single nucleotide polymorphism ($A_{-444}C$) described in the LTC_4 synthase gene promoter [10] has been postulated to enhance gene transcription and increase cellular capacity for cys-LT synthesis [6]. The variant LTC_4 synthase genotypes are reported to be highly prevalent (76%) in aspirin-intolerant asthma (AIA) patients [10], who have constitutively-elevated cys-LT synthesis and good clinical responses to anti-leukotriene therapy, and are also common (42-44%) in normal subjects and aspirin-tolerant asthma (ATA) patients. The results presented here provide preliminary evidence that, irrespective of aspirin sensitivity, the variant genotypes confer an increased capacity for cys-LT synthesis in eosinophils, the predominant cell-type expressing LTC_4 synthase in the asthmatic airway [4,5], and that in severe asthmatic patients, the variant genotypes may be markers of a leukotriene-dependent phenotype of asthma that responds well to therapy with a leukotriene receptor antagonist.

Differences in LTC_4 synthase expression in eosinophils and other leukocytes may only be rate-limiting for LTC_4 synthesis when the 5-LO pathway is highly stimulated. Thus, in eosinophils from normal subjects, there was no difference in LTC_4 synthesis *in vitro* between the variant and wild-type genotypes in the absence of ionophore stimulation, and low levels of leukocyte activation *in vivo* may also explain the lack of difference in urinary LTE_4 levels. However, stimulation of normal eosinophils with calcium ionophore revealed significantly elevated LTC_4 synthesis in subjects with the variant LTC_4 synthase genotype. The difference became statistically-significant in the presence of indomethacin, which prevents shunting of arachidonate along the cyclooxygenase pathway. This may explain why in most AIA patients exposure to aspirin and other COX inhibitors consistently provokes an acute surge of cys-LT synthesis leading to acute bronchoconstriction.

The increased prevalence of the variant LTC_4 synthase genotype (56%) in the severe asthmatics compared to the normal subjects (32%), and the concomitant increase in the capacity for LTC_4 production in eosinophils when stimulated, suggests that a genetic propensity for enhanced cys-LT production *in vivo* may be a predisposing factor for severe asthma irrespective of aspirin sensitivity. Only three of the 23 asthmatic patients had a clinical history of aspirin sensitivity.

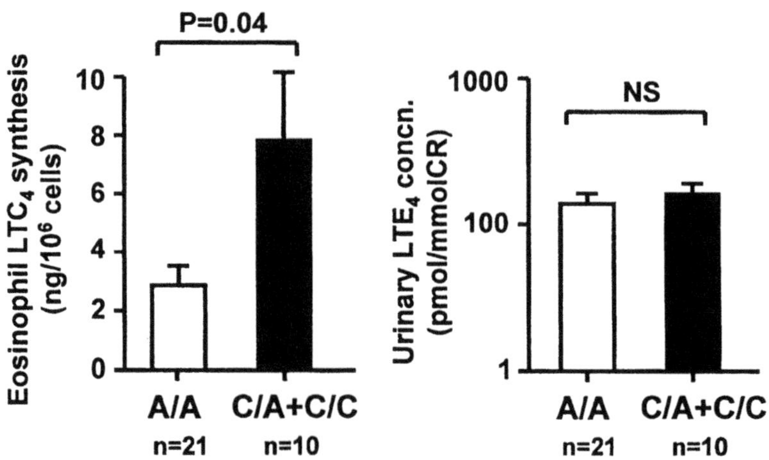

Figure 1. (Left) Release of immunoreactive LTC_4 from normal blood eosinophils stimulated for 15 minutes with calcium ionophore A23187 (1μM) in the presence of indomethacin (10μM). Eosinophils from normal subjects with variant LTC_4 synthase genotypes (C/C and C/A) generated more LTC_4 than those from subjects with the wild-type genotype (A/A). (Right) LTC_4 synthase genotype had no effect on basal levels of urinary LTE_4 in the same normal subjects.

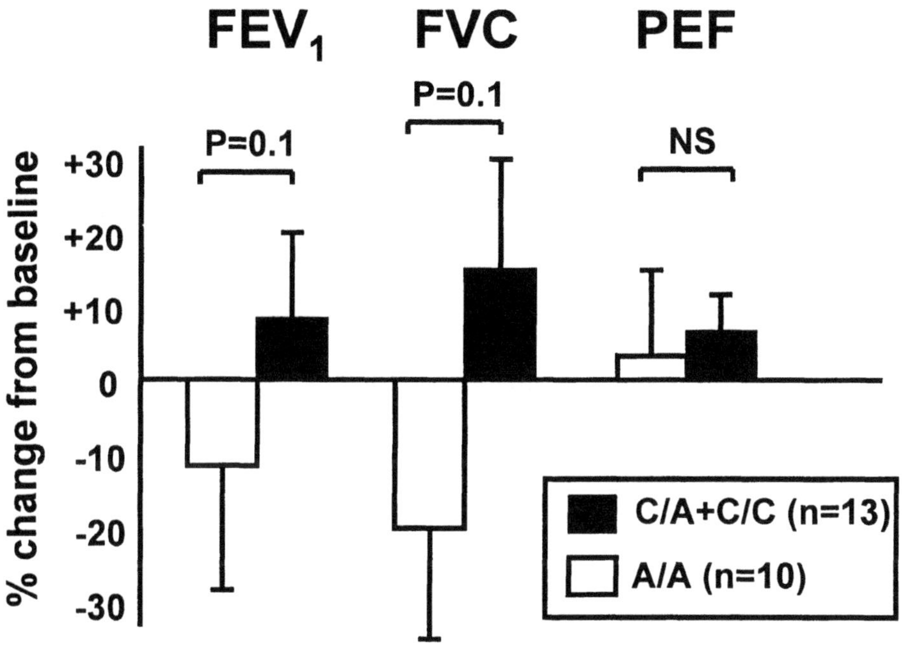

Figure 2. Percentage changes in lung function compared to pre-study values in severe persistent asthmatics following oral therapy with the cysteinyl-leukotriene receptor antagonist zafirlukast (Accolate; 20mg bd) for two weeks. Compared to patients with the wild-type LTC_4 synthase genotype (A/A; n=10), patients with variant genotypes (C/C and C/A; n=13) tended to show greater improvement in laboratory measures of forced expiratory volume in one second (FEV_1) and forced vital capacity (FVC)(both p=0.1), but not in home peak expiratory flow (PEF).

The study of two weeks of oral zafirlukast therapy in severe chronic asthmatics was an open study involving a small number of patients (n=23). Withdrawal of oral corticosteroids and other medications prior to the study probably accounts for the deteriorations in lung function observed in patients with the wild-type LTC_4 synthase genotype. Against this background, patients with the variant genotypes showed a tendency to improvements in two clinic measures of lung function (FEV_1 and FVC). Although not reaching statistical significance (P=0.1), the results suggest that enhanced cys-LT synthesis contributes disproportionately to asthma pathophysiology in the 56% of patients with the variant LTC_4 synthase genotype, and that these patients may represent the target group of good responders to anti-leukotriene therapy.

Acknowledgements

APS is funded by a grant-in-aid from Merck & Co., and IS by the National Asthma Campaign. STH is a Medical Research Council Clinical Professor (Programme Grant no. G8604034).

References

1. Hay DWP, Torphy TJ, Undem BJ. Cysteinyl leukotrienes in asthma: old mediators up to new tricks. Trends Pharmacol Sci 1995;16:304-9.
2. Drazen JM, Israel E, O'Byrne PM. Treatment of asthma with drugs modifying the leukotriene pathway. N Engl J Med 1999;340:197-206.
3. Sampson AP, Holgate ST. Leukotriene modifiers in the treatment of asthma. Br Med J 1998; 316:1257-58.
4. Cowburn A, Sladek K, Soja J, Adamek L, et al. Over-expression of leukotriene C_4 synthase in bronchial biopsies of aspirin-intolerant asthmatics. J Clin Invest 1998;101:834-46.
5. Seymour ML, Aberg D, Riise G, Rak S, Holgate ST, Sampson AP. Seasonal allergen exposure increases expression of leukotriene pathway enzymes and induces eosinophil influx in bronchial mucosa of atopic asthmatics. J Allergy Clin Immunol 1998;101:711 (Abstract).
6. Sanak M, Sampson AP. Biosynthesis of cysteinyl-leukotrienes in aspirin-intolerant asthma. Clin Exp Allergy 1999;29:306-13.
7. In K, Asano K, Beler D, et al. Naturally-ocurring mutations in the human 5-lipoxygenase gene promoter that modify transcription factor binding and reporter gene transcription. J Clin Invest 1997;99:1130-7.
8. Drazen JM, Yandava CN, Dube L, et al. Pharmacogenetic association between ALOX5 promoter genotype and the response to anti-asthma treatment. Nat Genet 1999;22:168-70.
9. Szczeklik A, Stevenson DD. Aspirin-induced asthma: advances in pathogenesis and management. J Allergy Clin Immunol 1999;104:5-13.
10. Sanak M, Simon HU, Szczeklik A. Leukotriene C_4 synthase promoter polymorphism and risk of aspirin-induced asthma. Lancet 1997;350:1599-1600.
11. Sanak M, Pierzchalska M, Bazan-Socha S, Szczeklik A. Enhanced expression of the leukotriene C_4 synthase due to overactive transcription of an allelic variant associated with aspirin-intolerant asthma. Am J Respir Cell Mol Biol 2000;23:290-96.
12. Van Sambeek R, Stevenson DD, Baldasaro M, et al. 5' flanking region polymorphism of the gene encoding leukotriene C_4 synthase does not correlate with the aspirin-intolerant asthma phenotype in the United States. J Allergy Clin Immunol 2000;106:72-76.
13. Cowburn AS, Holgate ST, Sampson AP. IL-5 increases expression of 5-lipoxygenase-activating protein and translocates 5-lipoxygenase to the nucleus in human blood eosinophils. J Immunol 1999;163:456-65.

ANTI-INFLAMMATORY LEVELS OF ASPIRIN AND SALICYLATE INHIBIT INTEGRIN-DEPENDENT HUMAN NEUTROPHIL AGGREGATION AND PMN INFLUX IN COX 2- AND NFκB (P105)-KNOCKOUT MICE

Gerald Weissmann, M. Carmen Montesinos, Michael Pillinger, and Bruce N. Cronstein

Introduction

The anti-inflammatory effects of aspirin have generally been attributed to direct inhibition of cyclooxygenases (COX-1 and COX-2) [1] but additional mechanisms are likely at work. These include aspirin's inhibition of NFκB translocation to the nucleus as well as the capacity of salicylates to uncouple oxidative phosphorylation (i.e. deplete ATP). Aspirin has been used for a century to reduce redness, swelling, heat, and pain; recently it has been shown to prevent intravascular thrombosis, slow Alzheimer's disease, and prevent colon cancer [2]. The effects of aspirin vary with dose, of which three levels are generally appreciated [2,3]. Low doses (80 mg/day; aspirin I) irreversibly acetylate serine 530 of cyclooxygenase (COX-1) to inhibit platelet thromboxane A_2 generation and are antithrombotic. Intermediate doses (2-4 g/day; aspirin II) globally inhibit COX-1 and 2, block prostaglandin production, and are analgesic and antipyretic. Finally, serum concentrations in the millimolar range (6-8 g/day; aspirin III) are as effective as cortisone in rheumatic disorders [2]. Added to human neutrophils *in vitro*, high concentrations of aspirin III or its active metabolite salicylate (NaS) inhibit homotypic cell adhesion and O_2.- generation, but NaS is only 1:100 as potent as aspirin at inhibiting the COXs [2,3].

Millimolar concentrations of salicylates also inhibit heterotypic adhesion of stimulated neutrophils to endothelial cells, a critical step in neutrophil-mediated injury [4,5]. High-dose salicylates also interfere with arachidonic acid (AA)-stimulated binding of GTP to preparations of neutrophil membranes [4]; decrease plasma membrane viscosity [6]; enhance intracellular cAMP levels and protein kinase A activity in neutrophils treated with FMLP [7]; and uncouple oxidative phosphorylation, thereby releasing micromolar amounts of adenosine, an autacoid with potent anti-inflammatory properties [8]. Salicylates also disrupt signal transduction between cell surface receptors and transcription of "inflammatory" cytokines by inhibiting the inhibitor of NFκB (IκBβ) kinase, thereby preventing translocation of NFkB to the nucleus [9]. Finally, Serhan's laboratory [10] has found that aspirin, but not sodium salicylate, promotes synthesis of such anti-inflammatory eicosanoids as 15-epi-lipoxin A4, while Vilcek's lab [11] has studied the mechanisms

B. Samuelsson et al. (eds.), Advances in Prostaglandin and Leukotriene Research, 149–154.

whereby salicylates promote apoptosis and inhibit TNF signalling via p38 mitogen-activated protein kinase.

Which of these *in vitro* effects of salicylates account for their anti-inflammatory action remains to be determined. Here we show two nonprostaglandin effects of salicylates on inflammation: 1) inhibition of Erk-dependent human neutrophil activation, and 2) inhibition of air-pouch inflammation in COX 2-and NFκB (p105)-knockout mice.

Salicylate Effects on Human Neutrophil Aggregation: Integrin/ Erk-Dependency.

The mitogen-activated protein kinases p44Erk1 and p42Erk2 are serine/threonine kinases that are activated by growth factors, etc. which act via kinases *ras,* Raf-1, and Mek [3,12]. Chemoattractants such as FMLP stimulate Erk activation in neutrophils at least in part by signals similar to those launched by protein tyrosine kinase receptors, i.e. via *ras,* Raf-1, and Mek [3]. Incubation of neutrophils with millimolar concentrations of aspirin or NaS inhibited Erk activation in response to FMLP or AA (Figure 1).

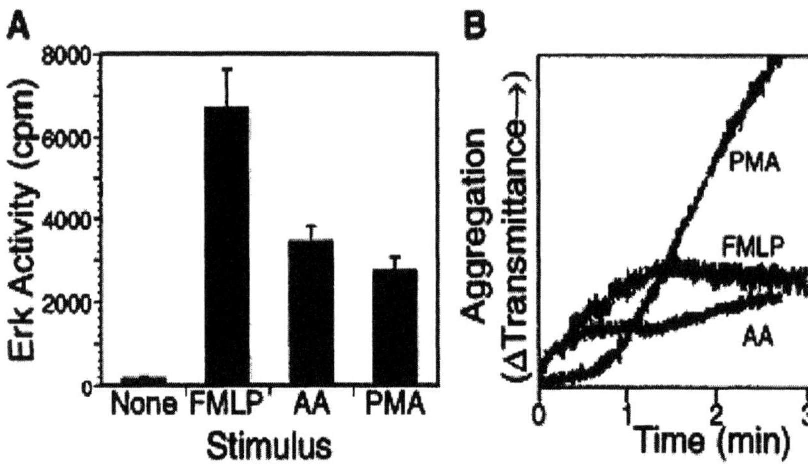

Figure 1. Stimulation of neutrophil Erk activity and homotypic aggregation. (A) Neutrophils stimulated for 2 min with FMLP (100 nM) or AA (20 μM) or PMA (10 min,1 μg/ml), and analyzed for Erk activity as described [3]. (B) Neutrophils stimulated and analyzed online for homotypic aggregation for 2 min as described [3]. Mean ± SEM of n=20 (A), or representative of 5 (B).

Micromolar aspirin concentrations sufficient for COX inhibition (i.e. aspirin I and II) had no effect on FMLP or AA stimulation of Erk. Indomethacin at doses sufficient for COX inhibition (1-10 μM) did not inhibit FMLP or AA-stimulated Erk activity [3]. Acetaminophen,to 3 mM, failed to affect Erk activity (Figures 2A and B). PMA bypasses membrane signaling to activate protein kinase C and aspirin III or NaS failed to inhibit PMA-stimulated Erk activation in neutrophils, excluding the possibility that salicylates exert direct effects on Erk, or on assays of Erk activation. Nor did acetaminophen influence PMA-stimulated Erk activation (Figure 2C)

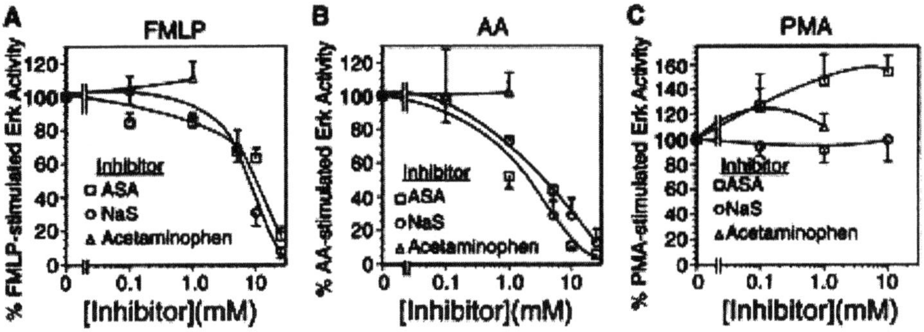

Figure 2. Effects of salicylates and acetaminophen on neutrophil Erk activity. Neutrophils incubated 10 minutes 37°C in the absence or presence of aspirin (ASA), NaS, or acetaminophen, followed by 2 minutes with 100 nM FMLP (A) or 20 μM AA (B), or 10 minutes 1 μg/ml PMA © at 37°C. Results expressed as percent of stimulated Erk activity without inhibitors; n=3, ± SEM.

We also tested the capacity of NaS or aspirin to inhibit Erk phosphorylation by immunoblotting with antiserum specific for phosphorylated, activated Erk. Neutrophil stimulation with 100 nM FMLP, with 20 μM AA (2 min each), or with 1 μg/ml of PMA (10 minutes) each caused dramatic increases in phosphorylation of Erk [3]. Incubation with 10 mM aspirin or NaS for 10 minutes before stimulation inhibited Erk phosphorylation in response to FMLP or AA but not PMA [3]. We next tested PD098059, a specific inhibitor of Mek (the proximal kinase activating Erk). PD098059 also inhibited Erk activation and aggregation in response to FMLP, as well as FMLP-stimulated neutrophil adhesion to ECV309 cell monolayers [3]. The data show that aspirin and NaS inhibit neutrophil Erk activation at anti-inflammatory concentrations that inhibit CD11b/CD18-dependent homotypic aggregation of neutrophils (i.e. aspirin III). The salicylate effect was upstream of Erk, and unlikely due to COX inhibition (as by aspirin I or II).

Salicylates are Anti-inflammatory in COX-2- and NFκB (P105)-Knockout Mice.

At clinically relevant doses, salicylates cause cells to release micromolar concentrations of adenosine, which serves as an endogenous ligand for at least four different types of well-characterized receptors. Previously, we have shown that adenosine mediates the anti-inflammatory effects of other potent and widely used anti-inflammatory agents, methotrexate and sulfasalazine, both *in vitro* and in *vivo* [13]. To determine *in vivo* whether clinically relevant levels (i.e. aspirin III or mM salicylate) act via adenosine, via NFκB, or via the "inflammatory" cyclooxygenase COX-2, we studied acute inflammation in the murine air-pouch model by using wild-type mice and mice rendered deficient in either COX-2 or p105, the precursor of p50, a component of multimeric NFκB. In some animals, an adenosine A2 receptor antagonist, DMPX (10 μM), or adenosine deaminase, ADA (0.125 units/ml) were injected into the air pouch with the inflammatory stimulus. ADA and DMPX significantly diminished the inhibition of leukocyte accumulation mediated by treatment with either aspirin or sodium salicylate ($P < 0.008$ for all; $n \geq 7$).

It is common wisdom that inducable COX-2 is responsible for the production of inflammatory eicosanoids [1], although recent data suggest that COX-1 may be critical for some inflammatory responses. We therefore determined whether aspirin retained its anti-inflammatory properties in animals with targeted disruption of the gene for COX-2 (see reference #20: B6;129-Ptgs2fm1Jed; The Jackson Laboratory). No difference was found in the response to aspirin between parental strain (wild-type; C57BL/6) animals and COX-2 knockouts (Figure3). The concentrations of prostaglandin E2 in the inflammatory exudates of COX-2 knockout mice were comparable with those found in controls, consistent with the observation that there is a compensatory increase of COX-1 activity in COX-2 knockout cells [14]. As expected, aspirin treatment reduced the concentration of prostaglandin E2 in the inflammatory exudates of COX-2 knockout mice [14].

Aspirin III and glucocorticoids are equally effective anti-inflammatory agents in acute rheumatism and in the murine air-pouch model [13], and these agents share at least one common therapeutic target: the activation and translocation of NFκB to the nucleus [2,9]. We therefore compared their effects in wild-type mice and in animals with targeted disruption of NFκB. Surprisingly disruption of NFκB p50 failed to affect the accumulation of leukocytes in air pouches of wild-type mice [$(2.6 \pm 0.2) \times 10^6$ cells per ml versus $(3.1 \pm 0.7) \times 10^6$ cells per ml; wild type versus knockout; $n = 8$; P = ns]. Nevertheless, aspirin, salicylate, and sulfasalazine retained their anti-inflammatory effects in the NFκB p50 knockouts, whereas, as expected, dexamethasone was significantly less anti-inflammatory (Figure 4). Indeed, aspirin inhibited leukocyte accumulation more completely in the NFκB knockout mice than the wild-type mice.

The dissociation between prostaglandin E2 concentration and inflammation indicates that neither aspirin nor sodium salicylate mediates their anti-inflammatory effects by inhibiting prostaglandin production and suggests that prostaglandin E2 is not the major mediator of inflammation in this model.

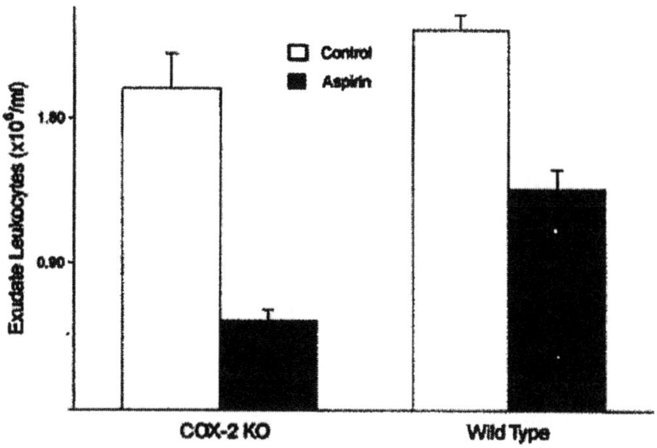

Figure 3. Aspirin is anti-inflammatory in animals rendered deficient in COX-2 activity [COX-2 knockouts (KO) see text]. Significantly fewer leukocytes were present in the air pouches of COX-2 knockout mice treated with aspirin ($P < 0.01$; $n = 5$) than in those of controls ($n = 4$) and in air pouches of wild-type mice ($P < 0.05$; $n = 5$) treated with aspirin versus controls ($P < 0.05$) Data ± SEM.

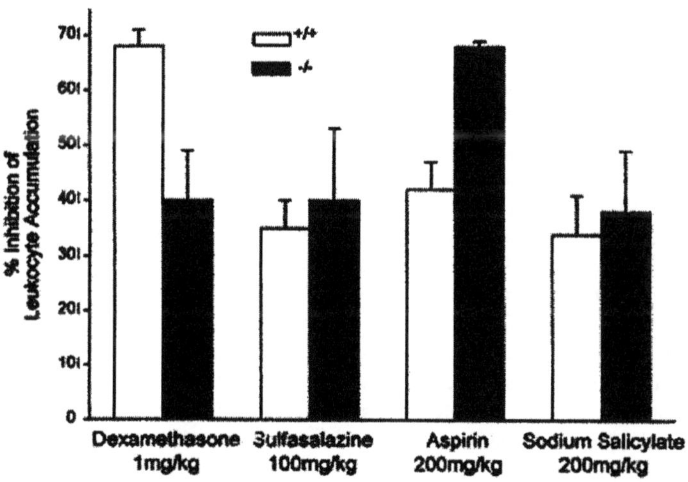

Figure 4. Aspirin, sodium salicylate, and sulfasalazine, unlike dexamethasone, are no less anti-inflammatory in p105 (NFκB) knockout mice than in wild-type mice. Animals were treated with aspirin or sodium salicylate as in Figure 3, dexamethasone as a single i.p. dose of dexamethasone 1 hour before inflammation, sulfasalazine by gastric gavage for 3 days before and on the day of the induction of inflammation. Leukocyte accumulation differed significantly between aspirin-treated and dexamethasone-treated knockout and wild-type mice ($P < 0.02$ for both, ± SEM, $n = 6\text{-}10$).

References

1. Vane JR. The mode of action of aspirin-like drugs. Nature (London) 1971;231:232-35.
2. Weissmann G. NSAIDs: aspirin and aspirin-like drugs. In: Wyngaarden JB, Smith, Jr. LH, Bennett JC, editors. Cecil Textbook of Medicine. Philadelphia: Saunders, 1992:114-18.
3. Pillinger MH, Capodici C, Rosenthal P, et al. Modes of action of aspirin-like drugs: salicylates inhibit erk activation and integrin-dependent neutrophil adhesion. Proc Natl Acad Sci USA 1998;95;14540-45.
4. Abramson SB, Leszczynska-Piziak J, Clancy RM, et al. Inhibition of neutrophil function by aspirin-like drugs (NSAIDS): requirement for assembly of heterotrimeric G proteins in bilayer phospholipid. Biochem Pharmacol 1994;47:563-72.
5. Cronstein BN, Van de Stouwe M, Druska L, Levin RI, Weissmann G. Nonsteroidal antiinflammatory agents inhibit stimulated neutrophil adhesion to endothelium: adenosine dependent andindependent mechanisms. Inflammation 1994;18:323-35.
6. Abramson SB, Leszczynska-Piziak J, Weissmann G. Arachidonic acid as a second messenger. Interactions with a GTP-binding protein of human neutrophils J Immunol 1991;147;231-36.
7. Abramson SB, Cherksey B, Gude D, et al. Nonsteroidal antiinflammatory drugs exert differential effects on neutrophil function and plasma membrane viscosity. in human neutrophils and liposomes. Inflammation 1990;4:11-30.
8. Cronstein BN. Adenosine: an endogenous antiinflammatory agent. In: Serhan CN, Ward PA, editors. Molecular and Cellular Basis of Inflammation. Totowa, NJ: Humana, 1998: 259-74.
9. Yin MJ, Yamamoto Y, Gaynor RB. The anti-inflammatory agents aspirin and salicylate inhibit the activity of I(kappa)B kinase-beta. Nature (London) 1998;396:77-80.
10. Chiang N, Fierro IM, Gronert K, Serhan CN. Activation of lipoxin A(4) receptors by aspirin-triggered lipoxins and select peptides evokes ligand-specific responses in inflammation. J Exper Med 2000;191:1197-1207.
11. Alpert D, Vilcek, J. Inhibition of IkappaB kinase activity by sodium salicylate in vitro does not reflect its inhibitory mechanism in intact cells. J Biol Chem 2000;275:10925-29.
12. Chardin P, Camonis JH, Gale W, et. al. Human Sos1: a guanine nucleotide exchange factor for Ras that binds to GRB2.. Science 1993;260:1338-43.
13. Cronstein BN, Naime D, Ostad E. The antiinflammatory mechanism of methotrexate. Increased adenosine release at inflamed sites diminishes leukocyteaccumulation in an in vivo model of inflammation. J Clin Invest 1993;92:2675-82.
14. Cronstein BN, Montesinos MC, Weissmann G. Salicylates and sulfasalazine, but not glucocorticoids, inhibit leukocyte accumulation by an adenosine-dependent mechanism that is independent of inhibition of prostaglandin synthesis and p105 of NFkappaB. Proc Natl Acad Sci USA 1999;96:6377-81.

Antipyretic Action of Acetaminophen: Inhibition of the Cyclooxygenase Activity of Endothelial Cells Treated with IL-1

John A. Oates, Lawrence J. Marnett, and Olivier Boutaud.

Introduction

The pharmacologic effects of acetaminophen are highly selective. Like the nonsteroidal anti-inflammatory drugs and inhibitors of prostaglandin-H-synthase-2 (PGHS-2), acetaminophen has antipyretic and moderate analgesic properties. Acetaminophen, however, exhibits little or none of the anti-inflammatory or antiplatelet effects nor the gastrointestinal toxicity produced by the nonsteroidal anti-inflammatory drugs. This research addresses the remarkable cellular selectivity in the inhibition of prostaglandin biosynthesis by acetaminophen.

An insight into the selectivity of acetaminophen derived from the findings of Flower and Vane [1] that the IC_{50} for inhibiting PGE_2 biosynthesis in the brain (93 μM) was within the range of clinical concentrations whereas in the spleen the IC_{50} was seven fold higher. It was inferred that blockade of prostaglandin biosynthesis resulted from inhibition of the cyclooxygenase, based on using exogenous arachidonic acid as substrate; these findings did not rule out an inhibition of PGE-synthase, however. The particular efficacy of acetaminophen in the nervous system has been confirmed by the finding that it is a potent inhibitor (IC_{50} 4.4 μM) of capsaicin-induced PGE_2 release from superfused spinal cord [2]. An additional locus of the selective action of acetaminophen was suggested by the findings of Green et al. [3] that acetaminophen inhibits prostacyclin biosynthesis in humans, a finding confirmed by O'Brien et al. [4] in a study of pregnant women. Analysis of the locus of prostacyclin synthesis by *in situ* hybridization of the mRNA encoding the prostacyclin synthase gene [5] and by other methods [6,7,8] indicates that in males, the major site of prostacyclin biosynthesis is the vessel wall, including both the smooth muscle and endothelial cells. That inhibition of prostacyclin biosynthesis by acetaminophen can occur in the endothelial cell was demonstrated by O'Brien [4]; in human umbilical vein endothelial cells (HUVEC) in culture they found 43% inhibition of prostacyclin biosynthesis by acetaminophen 66 μM but little further inhibition (47%) at 660 μM.

Results

Considering the context of acetaminophen's antipyretic action, we examined its effect on

B. Samuelsson et al. (eds.), Advances in Prostaglandin and Leukotriene Research, 155–158.

prostacyclin biosynthesis in HUVECs treated with interleukin-1α (IL-1α), and demonstrated an IC_{50} of 72 μM and 91% inhibition at 666 μM. Moreover, the IL-1α-induced increase in synthesis of prostacyclin made it possible to evaluate drug effect more efficiently, using smaller numbers of endothelial cells per sample in multi-well plates. Thus, the IL-1α-stimulated endothelial cells provide a relatively facile model in which to explore the factors that determine acetaminophen selectivity.

Because Green et al. [3] demonstrated an inhibition of the biosynthesis of PGI_2 but not that of thromboxane A_2 in humans, we addressed the question of whether the inhibition of PGI biosynthesis by acetaminophen resulted from inhibition of PGHS isozymes or inhibition of the PGI synthase. The activity of the PGI-synthase was determined in HUVEC, replicating the experimental system in which inhibition of PGI_2 biosynthesis by acetaminophen has been demonstrated previously [4]. [^{14}C]PGH_2 was synthesized to provide substrate for assessment of PGI synthase activity; hydrolysis of PGH_2 in aqueous solution yields PGE_2 and PGD_2, but no PGI_2 is formed in the absence of PGI synthase. Formation of 6-keto-$PGF_1\alpha$, the hydrolysis product of PGI_2, was measured as an indicator of PGI_2 biosynthesis. The formation of [^{14}C]6-keto-$PGF_1\alpha$ from [^{14}C]PGH_2 is not inhibited by concentrations of acetaminophen that markedly inhibit biosynthesis of PGI_2 in HUVEC. This finding, together with the fact that ApAP inhibits formation of PGI_2 from exogenous arachidonic acid, localized the action of ApAP to inhibition of one or both of the PGHS isozymes.

Prostaglandin E_2 (PGE_2), acting via the EP_3 receptor, is the prostaglandin mediator of fever [9]. It is therefore of interest that HUVECs make scant PGE_2 in the unstimulated state, but following IL-1α, we have demonstrated a marked increase (approximately 70 fold) in PGE_2 biosynthesis that exceeds the increment in PGI_2 formation. Acetaminophen inhibits the biosynthesis of PGE_2 in HUVEC in parallel with inhibition of prostacyclin biosynthesis. The induction of PGHS-2 in endothelial cells and inhibition of PGE_2 biosynthesis in these cells by acetaminophen implies that the endothelial cell could be the source of the PGE_2 that activates the hypothalamic thermoregulatory center. Indeed, the increase of PGHS-2 in the brain after IL-1 administration has been shown to occur in large part in endothelial cells by one group [10]. The endothelial cell locus of PGHS-2 induction was not detected, however, in previous investigations [11]. Accordingly, further evidence that localizes the site of PGHS-2 induction is required to address the attractive hypothesis that circulating pyrogens act on endothelial cells in the central nervous system to release the PGE_2 that initiates pyrexia.

The relatively potent effect of acetaminophen in HUVEC in which PGHS-2 is induced raised the question of whether acetaminophen inhibits PGHS-2 to a greater extent than PGHS-1. Accordingly, we determined the relative potency of acetaminophen on purified PGHS-1 (ovine) and PGHS-2 (murine) prepared from expression of the PGHS-2 gene in a baculovirus system. In this *in vitro* assessment, acetaminophen was not more potent as an inhibitor of PGHS-2 than of PGHS-1.

Because the inhibition of platelet aggregation and thromboxane A_2 (TxA_2) biosynthesis by acetaminophen is weak, the effect of the drug on TxA_2 biosynthesis by washed human platelet was investigated using the same arachidonic acid concentrations employed with the HUVEC studies (20 μM). There was little or no inhibition of platelet thromboxane A_2 production at 3 mM acetaminophen, quite in contrast with the effect of acetaminophen on PGI_2 biosynthesis by IL-1α stimulated HUVECs.

A second hypothesis for cellular selectivity in the response to acetaminophen is that the metabolic fate of the drug differs between cells in a way that could modify its efficacy through either formation of an active metabolite or by accelerated inactivation of the drug. There is abundant information on the biotransformation of acetaminophen and some of this does indeed indicate a tissue selectivity. In explorations of the mechanism of acetaminophen renal toxicity, deacetylation of acetaminophen to p-aminophenol (Figure 1) was demonstrated in the mouse kidney [12]. Subsequently, acetaminophen has been shown to undergo deacetylation to p-aminophenol followed by reacetylation back to acetaminophen, a "futile deacetylation" that confounds attempts to measure the extent of conversion of acetaminophen to p-aminophenol unless appropriate steps are taken to account for the futile deacetylation pathway [13]. Thus, it is of interest that acetaminophen is quite potent as an inhibitor of renal PGHS [14]. Further, inhibition of acetaminophen deacetylation reduces nephrotoxicity caused by acetaminophen but not that resulting from p-aminophenol, suggesting an important role of this metabolic pathway in nephrotoxicity [15]. Based on this and on evidence that p-aminophenol is a cosubstrate for PGHS-1, we evaluated the effects of p-aminophenol on PGHS.

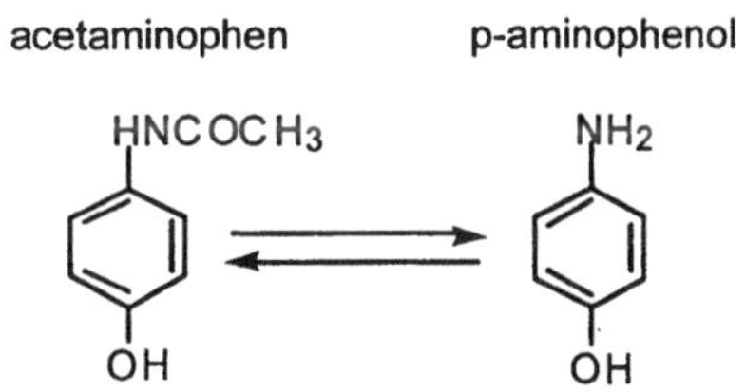

Figure 1. Metabolism of acetaminophen by deacetylation to p-aminophenol, which can be reacetylated back to acetaminophen.

We discovered that p-aminophenol is a much more potent inhibitor of PGHS-2 than is the parent drug. Using recombinant mouse PGHS-2, the oxygenation of [^{14}C] arachidonic acid was assessed by quantifying the radioactive oxygenation products with thin layer chromatography. In this vitro system the IC_{50} of acetaminophen was 3.1 mM whereas the IC_{50} of p-aminophenol was 0.25 mM. This suggests the possibility that in tissues in which acetaminophen is de-acetylated, the formation of p-aminophenol could lead to more potent inhibition of PGHS isozymes.

Acknowledgments

This work was supported in part by NIH Grant 15431, by a grant from Merck-Frosst, and by the Thomas F. Frist Sr. Chair in Medicine. The authors gratefully acknowledge the expert technical assistance of Elizabeth Shipp, Gwen Sobo, and Jemin Wang.

References

1. Flower RJ, Vane JR. Inhibition of prostaglandin synthetase in brain explains the anti-pyretic activity of paracetamol (4-acetamidophenol). Nature 1972;240:410-11.
2. Malmberg A, Yaksh T. Capsaicin-evoked prostaglandin E_2 release in spinal cord slices: relative effect of cyclooxygenase inhibitors. Eur J Pharmacol 1994;271:293-99.
3. Green K, Drvota V, Vesterqvist O. Pronounced reduction of in vivo prostacyclin synthesis in humans by acetaminophen (paracetamol). Prostaglandins 1989;37:311-15.
4. O'Brien WF, Krammer J, O'Leary TD, Mastrogiannis DS. The effect of acetaminophen on prostacyclin production in pregnant women. Presented at the Eleventh Annual Meeting of the American Gynecological and Obstetrical Society: Hot Springs, Virginia, September 10-12, 1992.
5. Tone Y, Inoue H, Hara S, et al. The regional distribution and cellular localization of mRNA encoding rat prostacyclin synthase. Eur J Cell Biol 1977;72:268-77.
6. Ager A, Gordon JL, Moncada S, Pearson JD, Salmon JA, Trevethick MA. Effect of isolation and culture on prostaglandin synthesis by porcine aortic endothelial and smooth muscle cells. J Cell Physiol 1982;110:9-16.
7. Baenzinger NL, Becherer PR, Majerus PW. Characterization of prostacyclin synthesis in cultured human arterial smooth muscle cells, venous endothelial cells and skin fibroblasts. Cell 1979;16:967-74.
8. Ingerman-Wojenski C, Silver MJ, Smith JB, Macarak E. Bovine endothelial cells in culture produce thromboxane as well as prostacyclin. J Clin Invest 1981;67:1292-96.
9. Ushikubi F, Segi E, Sugimoto Y, et al. Impaired febrile response in mice lacking the prostaglandin E receptor subtype EP3. Nature 1998;395:281-84.
10. Matsumura K, Cao C, Ozaki M, Morii H, Nakadate K, Watanabe Y. Brain endothelial cells express cyclooxygenase-2 during lipopolysaccharide-induced fever: light and electron microscopic immunocytochemical studies. J Neuroscience 1998;18:6279-89.
11. Elmquist JK, Breder CD, Sherin JE, et al. Intravenous lipopolysaccharide induces cyclooxygenase 2-like immunoreactivity in rat brain perivascular microglia and meningeal macrophages. J Comp Neurol 1997;381:119-29.
12. Carpenter HM, Mudge GH. Acetaminophen nephrotoxicity: studies on renal acetylation and deacetylation. J Pharmacol Exp Ther 1981;218:161-67.
13. Nicholls AW, Caddick S, Wilson ID, Farrant RD, Lindon JC, Nicholson JK. High resolution NMR spectroscopic studies on the metabolism and futile deacetylation of 4-hydroxyacetanilide (paracetamol) in the rat. Biochem Pharmacol 1995;49:1155-64.
14. Mattammal MB, Zenser TV, Brown WW, Herman CA, Davis BB. Mechanism of inhibition of renal prostaglandin production by acetaminophen. J Pharmacol Exp Ther 1979;210:405-9.
15. Newton JF, Kuo CH, DeShone GM, Hoefle D, Bernstein J, Hook JB. The role of p-aminophenol in acetaminophen-induced nephrotoxicity: effect of bis(p-nitrophenyl) phosphate on acetaminophen and p-aminophenol nephrotoxicity and metabolism in Fischer 344 rats. Toxicol Appl Pharmacol 1985;81:416-30.

Leukotriene-Mediated Alteration of Endothelial Barrier Function

Narinder Gautam, Lennart Lindbom, and Per Hedqvist

Introduction

In the host defense reaction to infection or tissue trauma inflammatory mediators act directly or indirectly on the endothelial cells (EC) to provoke increased permeability for macromolecules and extravasation of polymorphonuclear leukocytes (PMN). It is well established that enhanced macromolecular permeability *in vivo* in response to cysteinyl-leukotrienes (LTC_4-LTE_4), acting directly on the endothelium, is attributable to cell contraction and formation of interendothelial gaps [1]. To a limited extent, similar structural changes in EC shape may accompany PMN adhesion and migration across the endothelium in tissues stimulated with the chemotactic LTB_4 [2]. In this report, the kinetics of LT-induced changes in endothelial barrier function have been characterized, using confluent monolayers of bovine aorta and human umbilical EC, cultured on permeable membranes and mounted in a two-compartment diffusion chamber. This model permits continuous measurement of the electrical resistance across the monolayer (TEER) and analysis of transendothelial macromolecular efflux and PMN migration [3]. Our results indicate that LTC_4 acts directly on the EC, and LTB_4 via PMN activation and leukocytic CD18 adhesion-dependent mechanisms, to induce rapid rises in EC cytosolic [Ca^{2+}] and rearrangement of actin filaments that permit extravasation of macromolecules. Furthermore, LTC_4 and histamine in threshold concentrations act in concert to markedly enhance the structural changes in EC.

Results and Discussion

Addition of LTC4 (10^{-8} M) to the upper compartment of the flow chamber elicited a marked decrease in TEER, which was manifested within one minute, reached its maximum ($54 \pm 10\%$ of control) after 5 minutes (Figure 1), and remained at this level throughout the observation period (30 minutes). The fall in TEER was preceded by a rise in cytosolic [Ca^{2+}] and a marked increase in the number and density of filamentous actin stress fibers, and it was followed by an increase in macromolecular permeability. The responses to LTC_4 were dose-dependent, with LTC_4 (10^{-9} M) causing less pronounced changes in both EC resistance and albumin clearance and LTC_4 (10^{-10} M) apparently being ineffective. We have previously reported that histamine elicits closely similar changes in endothelium barrier

B. Samuelsson et al. (eds.), Advances in Prostaglandin and Leukotriene Research, 159–161.

function [3], with histamine (10^{-5} M) matching the effect of LTC_4 (10^{-8} M) on TEER and 10^{-7} M representing a threshold concentration in this respect (Figure 1). Coadministration of LTC_4 (10^{-10} M) and histamine (10^{-7} M), each *per se* without effect, elicited a marked decrease in TEER (45 ± 5% of control) (Figure 1) and increase in EC cytosolic [Ca^{2+}] and redistribution of actin filaments to the same extent as a 100-fold higher concentration of either LTC_4 or histamine. It has been reported that histamine may utilize intracellular calcium to modify cytoskeletal proteins in EC, and hence alter permeability [4]. Our results are in harmony with these data and indicate that LTC_4 and histamine jointly increase EC permeability for plasma proteins via transient rise in intracellular calcium and marked conformational changes in the EC cytoskeleton, in turn leading to cell contraction and opening of interendothelial junctions.

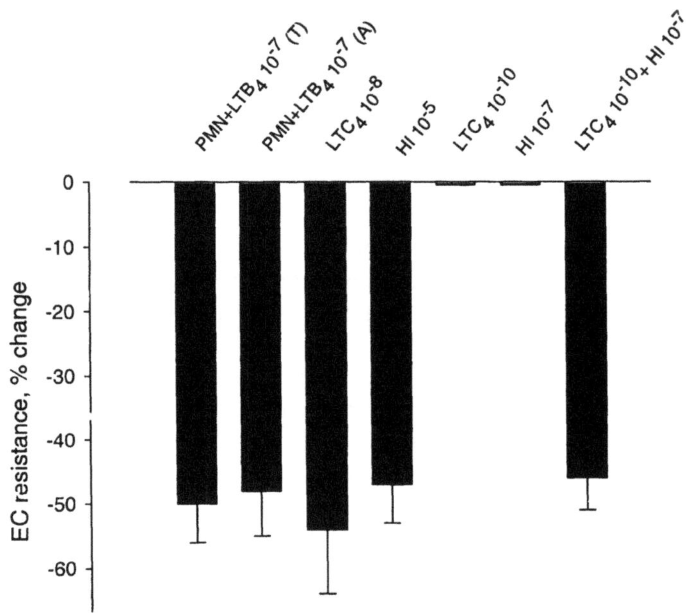

Figure 1. Magnitude of TEER changes in EC monolayers in response to transmural (T) or apical (A) LTB_4 stimulation of PMN resting in upper chamber, and in response to direct stimulation of EC with LTC_4 or histamine. Note dramatic fall in TEER after combined treatment with LTC_4 and histamine in threshold concentrations. Data are means ± SD based on 6-12 experiments in each group.

Transendothelial stimulation of PMN added to the upper compartment of the flow chamber and LTB_4 (10^{-7} M) added to the lower compartment 10 minutes later caused adhesion of PMN to the EC, and, in sequential order, a rise in EC cytosolic free Ca^{2+} and rearrangement of actin filaments, decrease in TEER, increased protein flux, and transmigration of PMN. The fall in TEER reached its maximum, 50 ± 6% of control (mean

± SD, n = 9), after 15 minutes (Figure 1), and then remained at this level throughout the observation period. The net albumin clearance started to rise approximately 5 minutes after start of challenge to a total of 18 ± 4 μl (mean ± SD) in 60 minutes. During the first 15 minutes of LTB4 challenge the transmigrated PMN were few in numbers (less than 1%) but then increased progressively, totaling 15 ± 3% (mean ± SD) of added PMN after 60 minutes.

Apical stimulation of PMN with LTB_4 (both PMN and LTB_4 added to the upper compartment) resulted in fall of TEER (48 ± 7% of control) and protein flux to the same extent as after transendothelial stimulation, but no transmigration of PMN. Adding PMN to the EC monolayer in the absence of LTB_4, or administration of LTB_4 alone, had no effect on TEER, indicating that an EC response requires the presence of both PMN and LTB_4. The kinetics of the EC and PMN responses to transendothelial and apical chemotactic stimulation with LTB_4 suggest that increase in EC permeability induced by activated PMN is unrelated to PMN transmigration. Both coherent [5] and conflicting [6] views with regard to dependency of protein leakage on PMN transmigration have been forwarded based on findings in other systems. We also found that the endothelial responses to chemotactic stimulation were dependent on physical interaction of PMN with EC via β2 integrins, inasmuch as anti-CD18 mAb treatment of the PMN abolished aii effects of LTB_4 challenge. Thus, PMN adhesion to, but not migration through the endothelium, is required for induction of the permeability change.

Acknowledgments

This study was supported by the Swedish Foundation for Health Care Sciences and Allergy Research (A98110), the Swedish MRC (14X-4342), and Karolinska Institute.

References

1. Joris I, Majno G, Corey EJ, Lewis RA. The mechanism of vascular leakage induced by leukotriene E_4. Endothelial contraction. Am J Pathol 1987;126(1):19-24.
2. Thureson-Klein A, Hedqvist P, Lindbom L. Leukocyte diapedesis and plasma extravasation after leukotriene B_4: lack of structural injury to the endothelium. Tissue Cell 1986;18(1):1-12.
3. Gautam N, Hedqvist P, Lindbom L. Kinetics of leukocyte-induced changes in endothelial barrier function. Br J Pharmacol 1998;125(5):1109-14.
4. Ehringer WD, Edwards MJ, Miller FN. Mechanisms of alpha-thrombin, histamine, and bradykinin induced endothelial permeability. J Cell Physiol 1996;167(3):562-69.
5. Huang AJ, Manning JE, Bandak TM, Ratau MC, Hanser KR, Silverstein SC. Endothelial cell cytosolic free calcium regulates neutrophil migration across monolayers of endothelial cells. J Cell Biol 1993;120(6):1371-80.
6. Rosengren S, Olofsson AM, von Andrian UH, Lundgren-Akerlund E, Arfors KE. Leukotriene B_4-induced neutrophil-mediated endothelial leakage in vitro and in vivo. J Appl Physiol 1991;71(4):1322-30.

GI and Cardiovascular Profiles of New NSAIDs: Selective COX-2 Inhibitors and NO-NSAIDs

Marcelo N. Muscará, Webb McKnight, Michael Dicay, and John L. Wallace

Introduction

In recent years, two new approaches have emerged for reducing the gastrointestinal toxicity of nonsteroidal anti-inflammatory drugs (NSAIDs); namely, selective inhibitors of cyclooxygenase (COX)-2 and nitric oxide-releasing NSAIDs (NO-NSAIDs). A large number of studies have been published which provide convincing data supporting the claims for gastrointestinal (GI)-sparing properties of these two classes of drugs [1-9]. In the case of the selective COX-2 inhibitors, the reduced GI toxicity is attributable to the sparing of GI prostaglandin (PG) synthesis, which is largely derived via COX-1. In the case of NO-NSAIDs, it is the slow release of nitric oxide (NO) that accounts for the reduced GI toxicity, most likely by maintaining gastrointestinal blood flow and inhibiting the activation of leukocytes within the GI microcirculation [5,6,10].

While these two classes of drugs share the feature of having minimal GI toxicity in healthy animals or humans, there are marked differences when one examines the effects of COX-2 inhibitors versus NO-NSAIDs in situations of pre-existing GI inflammation/ ulceration. For example, selective COX-2 inhibitors have been shown by several groups to delay the healing of gastric ulcers [11-13] and to exacerbate experimental colitis [14]. In contrast, NO-NSAIDs have been shown not to interfere with gastric ulcer healing [13], and in one study, to accelerate ulcer healing [15]. Moreover, NO-NSAIDs were found to be very well tolerated in a model of colitis in the rat [7] Marked differences between selective COX-2 inhibitors and NO-NSAIDs have also been demonstrated in models of ischemia-reperfusion injury in the GI tract. Whereas NO-NSAIDs can significantly reduce the gastric damage associated with hemorrhagic shock [16], selective COX-2 inhibitors were found to exacerbate the damage, presumably because COX-2 is up-regulated in response to ischemia in an effort to produce vasodilatory prostaglandins [17].

One must also consider that NSAIDs have significant toxicity outside of the GI tract; most notably in the kidney and in the cardiovascular system. Regarding the latter, NSAIDs can elevate blood pressure, at least in some individuals, and can interfere with the efficacy of antihypertensive medications [18-20]. It is noteworthy that studies performed by McAdam et al. [21] and Catella-Lawson et al. [22] demonstrated that two selective COX-2 inhibitors, celecoxib and rofecoxib, were able to markedly suppress systemic

B. Samuelsson et al. (eds.), Advances in Prostaglandin and Leukotriene Research, 163–169.

prostacyclin synthesis. This raised the possibility that selective blockade of COX-2, by virtue of inhibiting prostacylin synthesis but sparing thromboxane synthesis, might promote thrombosis. Moreover, in a recent study, we examined the possibility that a selective COX-2 inhibitor (celecoxib) would exacerbate pre-existing hypertension in rats [23]. Indeed, once-daily administration of celecoxib at an anti-inflammatory dose was shown to produce significant increases in systemic blood pressure in rats with hypertension induced by addition of a nitric oxide synthase inhibitor to the drinking water. More surprising, however, was the observation that this drug significantly and progressively increased blood pressure in normotensive rats. These effects appeared to be related to effects of celecoxib on renal function, which is consistent with clinical observations of fluid retention and hypertension in a significant percentage of patients taking selective COX-2 inhibitors.

In a recent, very large clinical trial, rofecoxib was found to be associated with significantly more cardiovascular adverse events (myocardial infarction, stroke) than the comparator NSAID, naproxen. This led to the suggestion that patients taking selective COX-2 inhibitors who may be at risk of such cardiovascular events should consider taking concomitant low-dose aspirin. However, little is known of the effects of the combined administration of a selective COX-2 inhibitor and low-dose aspirin on the gastric mucosa. Given the evidence that COX-2 inhibition can interfere with the healing of gastric ulcers in rats and mice [11,12] and can exacerbate ischemia-reperfusion associated gastric damage in rats [17], it is conceivable that the combination of aspirin with a selective COX-2 inhibitor could result in an increase in gastric damage. This possibility was examined in the present study. Moreover, we have assessed the possibility that a selective COX-2 inhibitor (celecoxib), as well as causing an elevation of blood pressure in hypertensive rats, may reduce the effectiveness of antihypertensive therapy in hypertensive rats. For comparison, we have also examined the effects of a nitric oxide-releasing derivative of aspirin (NCX-4016) in each of these models.

Effects of Celecoxib + Aspirin on the Gastric Mucosa

In previous studies, we determined the doses of celecoxib that were anti-inflammatory in the rat [23]. These studies involved the use of the carrageenan-airpouch model. At doses of celecoxib of 5 mg/kg or greater, significant suppression of COX-2 activity (inflammatory PGE_2 synthesis) was observed. We therefore tested the effects of intraperitoneal administration of celecoxib (10 mg/kg) on the gastric damaging effects of various doses of aspirin. Aspirin was given orally 30 minutes after celecoxib administration, and the severity of gastric damage was blindly scored 3 hours later.

As shown in Figure 1, aspirin caused dose-dependent increases in gastric damage. Administration of celecoxib alone did not cause significant gastric damage. However, prior administration of celecoxib resulted in a significant increase in the severity of aspirin-induced gastric damage at all doses of aspirin tested (Figure 1). In sharp contrast, the combination of celecoxib plus NCX-4016 did not result in detectable damage at any dose. Additional experiments were performed in which a fixed dose of aspirin (10 mg/kg) was given orally to rats, and a range of doses of celecoxib (intraperitoneal) were tested. As shown in Figure 2, celecoxib dose-dependently increased the severity of gastric damage induced by aspirin. It is noteworthy that celecoxib significantly increased gastric damage only at the doses previously shown to significantly inhibit COX-2 activity (i.e. > 5 mg/kg) [24]. Once again, oral administration of NCX-4016 together with intraperitoneal administration of celecoxib failed to elicit significant gastric damage.

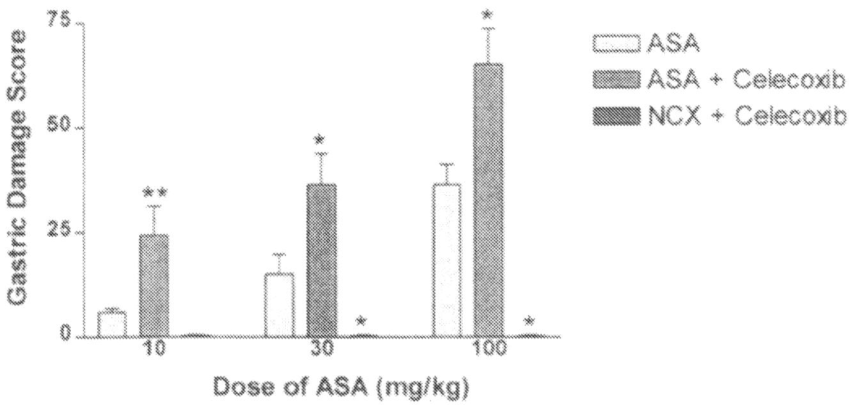

Figure 1: Effects of intraperitoneal celecoxib (10 mg/kg) plus orally administered aspirin or NCX-4016 (equimolar doses to those of aspirin) on severity of gastric damage. Results are shown as the mean ± SEM of at least 5 rats per group. *p < 0.05, **p < 0.01 compared to the group treated with aspirin alone.

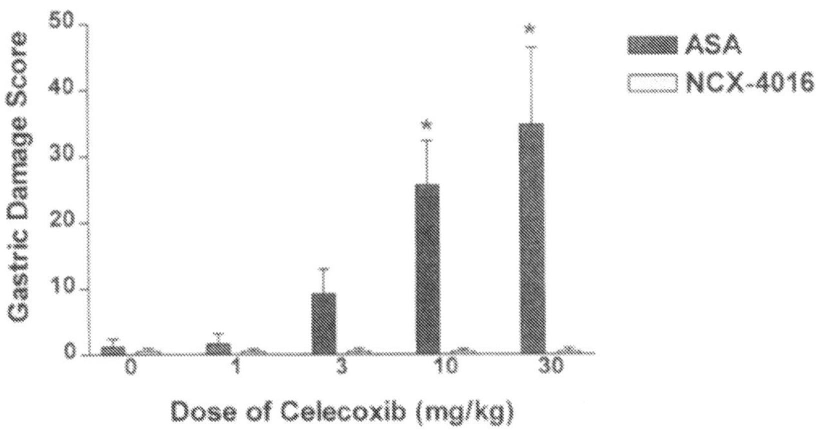

Figure 2: Effects of intraperitoneal celecoxib plus orally administered aspirin (10 mg/kg) or NCX-4016 (18.6 mg/kg; equimolar dose to the dose of aspirin) on severity of gastric damage. Results are shown as the mean ± SEM of at least 5 rats per group. *p < 0.05 compared to the group treated with aspirin or NCX-4016 alone.

The increased gastric toxicity of the combination of a selective COX-2 inhibitor with a conventional NSAID is consistent with our recent finding that it is necessary to inhibit both COX-1 and COX-2 in order to elicit gastric damage with an NSAID [25]. In that study, the combination of a selective COX-2 inhibitor and a selective COX-1 inhibitor consistently produced gastric injury, while either drug alone did not.

Effects of Celecoxib versus NO-NSAID on the Antihypertensive Effects of Enalapril

Hypertension was induced in rats by addition to their drinking water of the nitric oxide synthase inhibitor, L-NAME, as described previously [23]. L-NAME treatment resulted in a rapid increase in systemic arterial blood pressure (Figure 3). The mean systolic blood pressure had increased by ~40 mm Hg within two weeks of beginning the treatment with L-NAME. The systolic blood pressure continued to increase in the group treated with vehicle each day, reaching a mean of 163 mm Hg. Treatment with enalapril each day resulted in a rapid and significant reduction of blood pressure, reaching a plateau level of ~135 mm Hg. However, concomitant administration of celecoxib (10 mg/kg) interfered with the efficacy of enalapril, such that the blood pressure in those rats was not significantly different from that in rats treated with vehicle. On the other hand, concomitant administration of NCX-4016 (18.6 mg/kg, equimolar to 10 mg/kg aspirin) and enalapril did not significantly affect the ability of enalapril to reduce blood pressure.

Conclusions

The results presented herein demonstrate that administration of a selective COX-2 inhibitor (celecoxib) can result in significant exacerbation of aspirin-induced damage in the rat stomach. The exacerbation of damage by celecoxib occurs in a dose-dependent manner. In contrast, the NO-releasing aspirin derivative, NCX-4016, did not cause detectable gastric damage when given at doses equimolar to the doses of aspirin that were tested, nor was damage seen when NCX-4016 was given together with celecoxib. Thus, it is likely that the NO released by NCX-4016 is able to protect the gastric mucosa from the damage that would normally be produced by combined inhibition of COX-1 and COX-2.

Celecoxib administration at an anti-inflammatory dose interfered with the antihypertensive effects of enalapril. In contrast, NCX-4016 did not exhibit such effects. We have previously demonstrated that celecoxib significantly exacerbated pre-existing hypertension in the rat, and significantly elevated blood pressure in normal rats [23]. It is possible that these effects of celecoxib are specific to the rat, although hypertension was observed in a significant percentage of patients in several clinical trials of selective COX-2 inhibitors. Thus, our results would suggest that NO-NSAIDs represent an attractive alternative to selective COX-2 inhibitors for treatment of inflammatory conditions in patients with pre-existing hypertension and/or at risk of other cardiovascular disease, as well as in patients with pre-existing gastrointestinal inflammation.

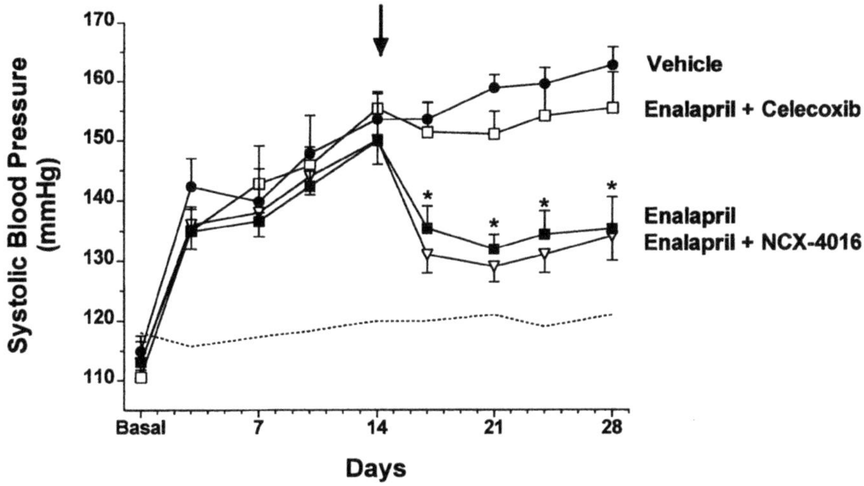

Figure 3: Effects of daily treatment with enalapril (25 mg/kg) ± celecoxib (10 mg/kg) or ± NCX-4016 (18.6 mg/kg) on systolic blood pressure in rats with hypertension induced by addition to the drinking water of a nitric synthase inhibitor (L-NAME; 400 mg/L). Treatment with the drugs or vehicle was started on day 14 of L-NAME ingestion (arrow), at which time hypertension was established. The dotted line shows the mean blood pressure in rats not receiving L-NAME. Note that the antihypertensive effects of enalapril were abolished by concomitant administration of the selective COX-2 inhibitor (celecoxib), but not affected by concomitant administration of NCX-4016. $*p < 0.05$ comparing the enalapril group or the enalapril + NCX-4016 group to the other groups. Each group consisted of 5 rats.

Acknowledgments

Dr. Muscará is supported by a fellowship from Merck & Co. Dr. Wallace is a Medical Research Council of Canada Senior Scientist and an Alberta Heritage Foundation for Medical Research Senior Scientist.

References

1. Boyce S, Chan CC, Gordon R, et al. L-745,337: a selective inhibitor of cyclooxygenase-2 elicits antinociception but not gastric ulceration in rats. Neuropharmacology 1994;33: 1609-11.
2. Masferrer JL, Zweifel BS, Manning PT, et al. Selective inhibition of inducible cyclooxygenase 2 in vivo is antiinflammatory and nonulcerogenic. Proc Natl Acad Sci USA 1994;91:3228-32.

3. Seibert, K., Zhang, Y., Leahy, K., et al. Pharmacological and biochemical demonstration of the role of cyclooxygenase 2 in inflammation and pain. Proc Natl Acad Sci USA 1994; 91:12013-17.
4. Chan CC, Boyce S, Brideau C, et al. Pharmacology of a selective cyclooxygenase-2 inhibitor, L-745,337: a novel nonsteroidal anti-inflammatory agent with an ulcerogenic sparing effect in rat and nonhuman primate stomach. J Pharmacol Exp Ther 1995;274: 1531-37.
5. Wallace JL, Reuter B, Cicala C, McKnight W, Grisham MB, Cirino G. Novel NSAID derivatives with markedly reduced ulcerogenic properties. Gastroenterology 1994;107: 173-79.
6. Wallace JL, Reuter B, Cicala C, McKnight W, Grisham MB, Cirino G. A diclofenac derivative without ulcerogenic properties. Eur J Pharmacol 1994;257:249-55.
7. Reuter BK, Cirino G, Wallace JL. Markedly reduced intestinal toxicity of a diclofenac derivative. Life Sci 1994;55:PL1-PL8.
8. Wallace JL, McKnight W, Del Soldato P, Baydoun AR, Cirino G. Anti-thrombotic effects of a nitric oxide-generating, gastric-sparing aspirin derivative. J Clin Invest 1995;96:2711-18.
9. Davies NM, Roseth AG, Appleyard CB, et al. NO-naproxen vs. naproxen: ulcerogenic, analgesic and anti-inflammatory effects. Aliment Pharmacol Ther 1997;11:69-79.
10. Wallace, JL, Elliott SN, Del Soldato P, McKnight W, Sannicolo F, Cirino G. Gastrointestinal sparing anti-inflammatory drugs: the development of nitric oxide-relasing NSAIDs Drug Develop Res 1997;42:144-49.
11. Mizuno H., Sakamoto C, Matsuda K, et al. Induction of cyclooxygenase 2 in gastric mucosal lesions and its inhibition by the specific antagonist delays healing in mice. Gastroenterology 1997;112:387-97.
12. Schmassmann A, Peskar BM, Stettler C,et al. Effects of inhibition of prostaglandin endoperoxide synthase-2 in chronic gastro-intestinal ulcer models in rats. Br J Pharmacol 1998;123:795-804.
13. Takeuchi K, Suzuki K, Yamamoto H, Araki H, Mizoguchi H, Ukawa H. Cyclooxygenase-2 selective and nitric oxide-releasing nonsteroidal anti-inflammatory drugs and gastric mucosal responses. J Physiol Pharmacol 1998;49:501-13.
14. Reuter BK, Asfaha S, Buret A, Sharkey KA, Wallace JL. Exacerbation of inflammation-associated colonic injury in rat through inhibition of cyclooxygenase-2. J Clin Invest 1996; 98:2076-85.
15. Elliott SN, McKnight W, Cirino G, Wallace JL. A nitric oxide-releasing nonsteroidal anti-inflammatory drug accelerates gastric ulcer healing in the rat. Gastroenterology 1995;109: 524-30.
16. Wallace JL, McKnight W, Wilson TL, Del Soldato P, Cirino G. Reduction of shock-induced gastric damage by a nitric oxide-releasing aspirin derivative: role of neutrophils. Am J Physiol 1997;273:G1246-G51.
17. Maricic N, Ehrlich K, Gretzer B, Schuligoi R, Respondek M, Peskar BM. Selective cyclo-oxygenase-2 inhibitors aggravate ischaemia-reperfusion injury in the rat stomach. Br J Pharmacol 1999;128:1659-66.
18. Fujihara CK, Malheiros DMAC, Donato JL, Poli A, de Nucci G, Zatz R. Nitroflurbiprofen, a new nonsteroidal anti-inflammatory, ameliorates structural injury in the remnant kidney. Am J Physiol 1998;274:F573-F79.

19. Goldblatt H, Haas E, Klick RL, Lewis LV. The effect of main artery occlusion of one kidney on blood pressure of dogs. Proc Natl Acad Sci U S A 1976;73:1722-24.
20. Whelton A. Nephrotoxicity of nonsteroidal anti-inflammatory drugs: physiologic foundations and clinical implications. Am J Med 1999;106:13S-24S.
21. McAdam BF, Catella-Lawson F, Mardini IA, Kapoor S, Lawson JA, FitzGerald GA. Systemic biosynthesis of prostacyclin by cyclooxygenase (COX)-2: the human pharmacology of a selective inhibitor of COX-2. Proc Natl Acad Sci USA 1999;96:272-77.
22. Catella-Lawson F, McAdam B, Morrison BW, et al Effects of specific inhibition of cyclooxygenase-2 on sodium balance, hemodynamics, and vasoactive eicosanoids. J Pharmacol Exp Ther 1999;289:735-41.
23. Muscará MN, Vergnolle N, Lovren F, Triggle DR, Elliott SN, Asfaha Wallace JL. Selective cyclooxygenase-2 inhibition with celecoxib elevates blood pressure and promotes leukocyte adherence. Br J Pharmacol 2000:129:1423-1430.
24. Wallace JL, Chapman K, McKnight W. Limited anti-inflammatory efficacy of cyclooxygenase-2 inhibition in carrageenan-airpouch inflammation. Br J Pharmacol 1999; 126:1200-1204.
25. Wallace JL, McKnight W, Reuter BK, Wallace JL. NSAID-induced gastric damage in the rat: requirement for inhibition of both cyclooxygenase-1 and -2. Gastroenterology, 2000; in press.

NO-RELEASING NSAIDS MODULATE CYTOKINE SECRETION

Stefano Fiorucci and Elisabetta Antonelli

IL-1β–converting enzyme (ICE, caspase-1) is the prototype of a family of cysteine proteases termed caspases, sharing the active site cysteine and aspartate binding clefts. The biological function of caspase-1 was originally thought to be restricted to the maturation process of IL-1β, a central mediator in the cytokine network . Proteolytic maturation of the inactive, 33-kD IL-1β precursor (proIL-1β) into the 17-kD, biologically functional form results from cleavage at the Asp^{116}-Ala^{117} site [1,2]. However, recent studies revealed that caspase-1 mediates processes relevant to immune and inflammatory reactions, in addition to IL-1β activation. The enzyme has been implicated in the activation of another inflammatory cytokine, IL-18, originally termed IFN-γ inducing factor (IGIF). Similar to IL-1β, IL-18 is expressed as an inactive precursor (24 kD), requiring proteolytic conversion via caspase-1 into the active, 18-kD form. The biological function of IL-18 was originally thought to be restricted to the induction of another inflammatory cytokine, IFN-, which has well-known roles in inflammation and Th1 immune responses. However, recent work indicated a much broader role for IL-18 in immunity and inflammation, mediating the expression of other proinflammatory cytokines, certain chemokines, and Fas ligand. In addition to the activation of proinflammatory mediators, caspase-1 is the prototype of a family of enzymes implicated in the control of apoptosis. Although the exact function of caspase-1 is controversial, this enzyme may participate in programmed cell death. In recent years ICE has been recognized as a member of the growing family of intracellular cysteine proteases that share sequence homology with *Ced-3*, a nematode gene involved in the execution phase of apoptosis [3-5]. The mammalian counterpart of the *Ced-3* gene products includes at least 14 different endoproteases that have been renamed caspases to denote cysteine proteases acting after an aspartic acid residue. Caspase-1, which has the greatest specificity for cleaving proIL-β, denotes the original ICE (both terms are often used interchangeably).

Comparison of molecular structures suggests that the caspase family falls into three major groups: caspases that function primarily in cytokine maturation (e.g. caspase-1, 4, and 5); initiator caspases, involved in signaling early steps of extracellular regulated apoptosis (e.g. caspase-8, 9, and 10); and, effector proteases involved in the execution phase of apoptosis (e.g. caspase-3, 6, and 7). Supporting the specialization of the three branches of caspase family, specific ICE inhibitors administered to mice exert poor anti-apoptotic effects although they reduce inflammation as effectively as does blocking IL-1β activity with specific antagonists [6].

B. Samuelsson et al. (eds.), Advances in Prostaglandin and Leukotriene Research, 171–178.

Although a caspase inhibitor has yet reached the clinic for the inhibition of apoptosis (ICE inhibitors have done so recently for the treatment of rheumatoid arthritis), preclinical studies are compelling, and the plethora of diseases in which they might have efficacy makes their therapeutic potential enormous. With few exceptions, all of the proof-of-concept preclinical studies with caspase inhibitors in animal models of human diseases have been performed with active-site mimetic peptide ketones (for example, benzyloxycarbonyl (z)-VAD-fluoromethylketone (fmk), z-YVAD-fmk/chloromethylketone (cmk), z-DEVD-fmk/cmk and z-D-cmk). These molecules are all relatively nonselective caspase inhibitors and although they are not appropriate tools for dissecting out the contributions made by individual caspase family members to the apoptotic response that occurs in disease models, they have provided extremely valuable preclinical insight into the potential that caspase inhibition might eventually have in humans. For example, in at least five different models of ischaemia–reperfusion injury (liver, cardiac, renal, intestinal, and cerebral), the caspase inhibition has shown remarkable efficacy. In addition to decreased apoptosis, caspase inhibition improved survival (for example threefold in liver ischemia), decreased infarct volumes (by 50% in both cardiac and focal cerebral ischemia) and, ultimately, markedly improved organ function (for example renal function and neurodeficit in models of kidney and focal cerebral ischaemia, respectively .

Nitric oxide (NO) releasing nonsteroidal anti-inflammatory drugs (NO-NSAIDs), are a recently described class of NSAID derivatives generated by adding an nitroxybutyl moiety through an ether linkage to the parental NSAID [7-11]. These compounds exhibit a markedly reduced gastrointestinal toxicity, while retaining the anti-inflammatory and antipyretic activity of parent NSAID. Although NO-NSAIDs spare the gastric mucosa, they inhibit prostaglandin generation and exert powerful antiapoptotic and anti-inflammatory effects [12]. Indeed, preliminary animal studies indicate that NO-NSAIDs are more effective than conventional NSAIDs in reducing inflammation and pain in arthritic rats. Previous studies from our laboratory have demonstrated that, similarly to conventional NO donors, NO-releasing NSAIDs inhibit apoptotic pathways by causing the S-nitrosylation/inactivation of effector caspases, such as caspase 3 [13].

NO-Aspirin Inhibits LPS-Stimulated Cytokine Release from Monocytes/Macrophages

It has been known, in animal models and in humans, that the bacterial endotoxin [LPS] induces an acute immune response inducing the secretion of different cytokine such as IL-1β, IL-2, IL-8, TNFα IL-18, and IFNγ, which are involved in the development of inflammation.

As shown in Figure 1, panel A, the incubation of human PMBC-derived monocytes with LPS for 24 hours resulted in a concentration-dependent stimulation of IL-1β and IL-18 release in cell supernatants with a peak occurring at 1 μg/ml). Moreover, as shown in Figure 1, panel C and D, the exposure of PMBC-derived monocytes to an NO-aspirin derivative, resulted in a concentration-dependent inhibition of IL-1β and IL-18 release induced by 1 μg/ml LPS. The inhibition of both cytokines release was significant at 10 μM ($P < 0.05$ for both cytokines), reached the half-maximum at 21 ± 6.3 and 25 ± 5.1 μM, respectively, and plateau at 100 μM (Figure 1, panel C and D). In contrast to NCX-4016, equimolar concentrations of aspirin failed to inhibit IL-1β and IL-18 release.

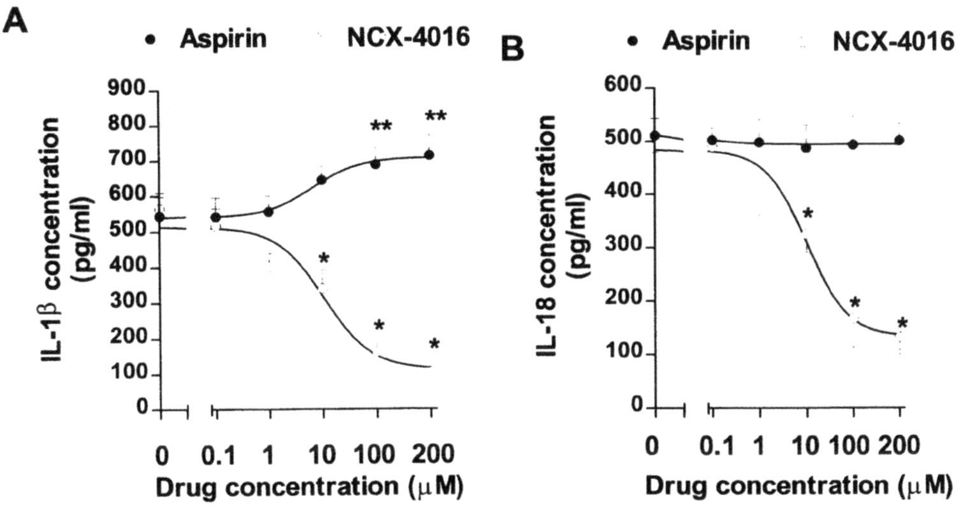

Figure 1. Panel A and B. NO-aspirin inhibits IL-1β and IL-18 release from LPS-challenged monocytes. Monocytes were stimulated with LPS (1μg/ml) in the presence or absence of increasing concentrations of NO-asprin or aspirin for 24 hours. The data are mean ± SE of 6 donors. *P < 0.001 versus cells incubated with the medium alone. **P < 0.01 versus cells incubated with the LPS alone.

NCX-4016 Inhibits ICE Activity Via S-Nitrosylation

To demonstrate the possible anti-inflammatory mechanism exerted by NO-aspirin we have first assessed that NO-aspirin inhibits cytokines release through a COX-independent mechanism and than the IL-1β and IL-18 release from LPS-stimulated human monocytes is regulated by an ICE-dependent pathway.

As illustrated in Figure 2, panel A, incubating monocytes with NO-aspirin, but not with aspirin, resulted in a concentration-dependent inhibition of ICE-like proteases with an IC_{50} of 18.3 ± 5.6 μM (Figure 2, panel B). Maximal inhibition of 78.2 ± 6.4% was observed at the concentration of 100 μ of NO-aspirin.

The S-nitrosylation is a well-established mechanism, by which NO inhibits enzyme activity [14-20] and as shown in Figure 3, the incubation of PMBC-derived monocytes with 100 μM SNAP and NO-aspirin, but not with aspirin, inhibited ICE-like cytokine protease activity, measured by assessing YVAD.AMC peptide cleavage. Incubating monocyte lysates with 20 mM DTT to remove thiol-bound group from proteins reverted the inhibitory effect exerted by the two NO-releasing agents, resulting in ≈65% recover of protease activity.

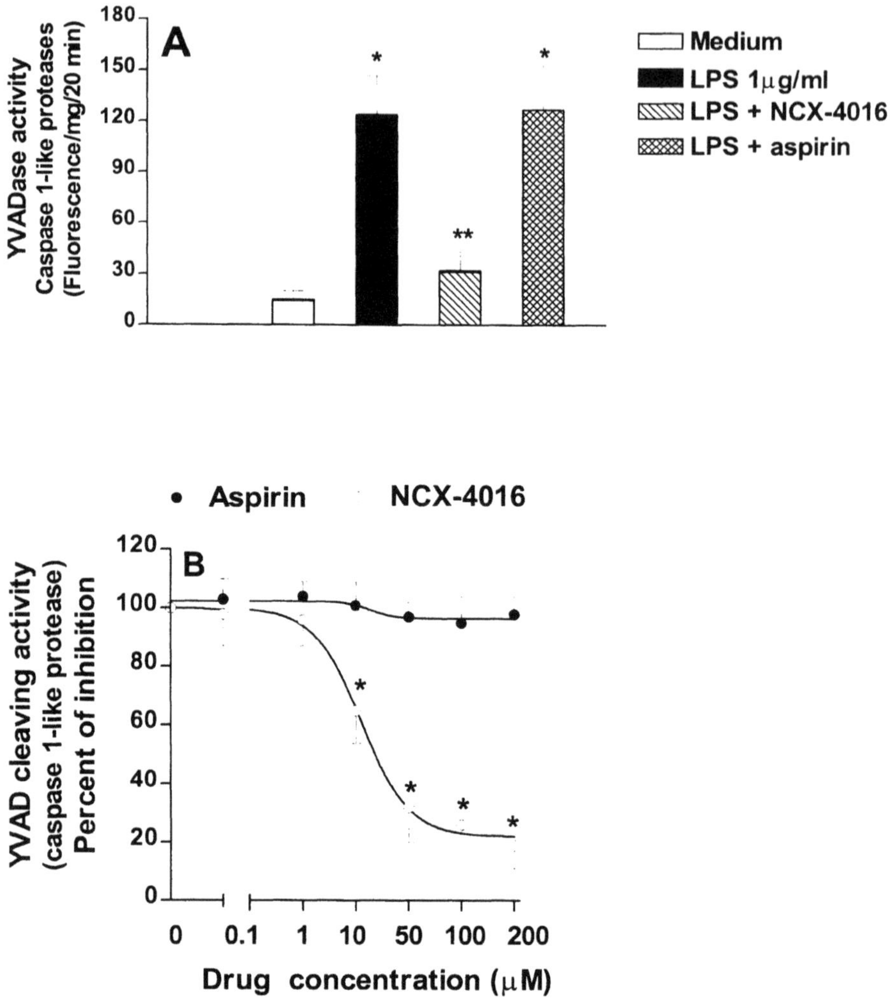

Figure 2. Panel A. NO-Aspirin inhibits ICE activation induced by LPS. YVADase activity was measured in lysates obtained from PMBC-derived monocytes stimulated with LPS, 1 μg/ml, in the presence or absence of 100 μM NO-aspirin or aspirin for 24 hours. The data are mean ± SE of 6 donors. * $P < 0.01$ versus cells incubated with medium alone. **$P < 0.01$ versus cells incubated with the LPS alone. Panel B. YVADase inhibition by NO-aspirin is concentration-dependent.

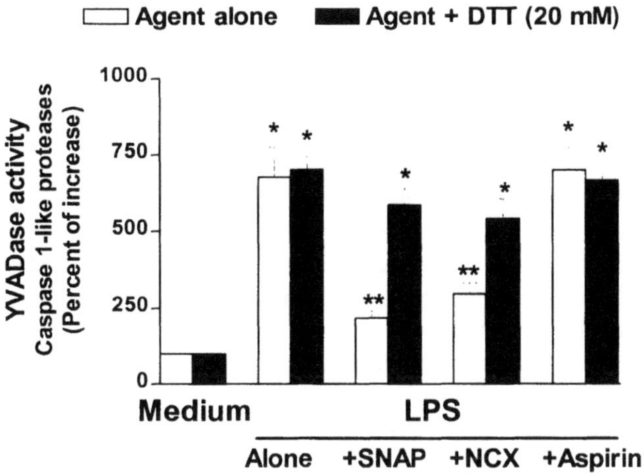

Figure 3. NO-Aspirin causes ICE-like proteases S-nitrosation/inhibition. Reversibility of inhibition of ICE-like cysteine proteases caused by SNAP and NO-aspirin by DTT. Lysates obtained from monocytes pretreated with LPS alone or in combination with SNAP, NO-aspirin or aspirin were incubated with 20 mM DTT and ICE activity measured. Data are mean ± SE of 6 determinations carried out in triplicate. *$P < 0.01$ versus homogenates incubated with medium alone. **$P < 0.01$ in comparison with homogenates incubated without DTT.

Discussion

Increasing evidence suggests a critical role of caspase-1 in inflammation, linking enhanced caspase-1 activity with the progression of various diseases, e.g., via the activation of the proinflammatory cytokine IL-1β and furthermore by promoting apoptosis. The emerging role of caspase-1 in the pathogenesis of inflammation had lead to the development of potent specific inhibitor that are able to cross the membrane of cells . Some of these compounds are currently investigated by various pharmaceutical companies (Merck, SmithKline Beecham, BASF, Idun/Novartis, and Vertex) although very little is known about their preclinical status or clinical agendas [12]. It is also clear that the first entries will probably target acute apoptotic injuries such as cerebral stroke, trauma-induced neurodegeneration, cardiac ischemia–reperfusion injury, transplantation, acute liver injury, and sepsis. In this context the ability of NO-aspirin to inhibit caspase-1 activity shows promise for treatment of caspase-mediated disease.

NO-releasing NSAIDs are a recently described class of NSAID derivatives generated by adding a nitroxybuthyl moiety through an ether linkage to the parental NSAID [7-11]. These compounds are virtually devoid of gastrointestinal toxicity, while retaining anti-inflammatory, antithrombotic, and antipyretic activity. Moreover, in some experimental models, these "NO-NSAIDs" exhibited significantly enhanced anti-inflammatory and analgesic effects in comparison with their parental molecules.

Moreover, we have provided the first evidence that the addition of an NO-releasing moiety to aspirin, confers new pharmacological properties to this molecule [10,11]. Indeed, we demonstrated that NO-aspirin inhibits pro-inflammatory cytokine release from LPS-challenged monocytes through a mechanism that involves the S-nitrosylation/inhibition of proteases required for cellular processing/maturation of IL-1β and IL-18.

Support for this view comes from following results: first, incubating monocytes with LPS resulted in a time-dependent activation of ICE-like endoproteases, an effect that was almost completely prevented by co-treating cells with NO-aspirin, but not with aspirin and that was unrelated to its ability to inhibit cyclo-oxygenase activity; second, inhibition of LPS-induced YVAD cleaving activities by NO-aspirin was reverted by DTT, an agent that effectively removes thiol-bound NO groups from proteins [14-17]

Taken together these data demonstrated that incubating human monocytes with NO-aspirin results in intracellular NO formation and S-nitrosation/inhibition of ICE-like cysteine proteases involved in pro-IL-1β and pro-IL-18 processing [18-23].

NO has previously been found to play a role in inflammation [24,25]. In addition to the well-established pro-inflammatory effect, it is now well recognized that NO acts as a double-edge sword being anti-inflammatory at low (micromolar) concentrations, and cytotoxic and pro-inflammatory at high (millimolar) concentrations [26]. Although the precise mechanisms by which endogenous NO exerts these effects are only partially known, a growing body of evidence indicates that low levels NO cause the S-nitrosylation of thiol groups located in the catalytic core of cysteine proteases required for cellular processing of pro-inflammatory cytokine [9,11,24]. Although the reversal of the NO-mediated inhibition by DTT is consistent with the S-nitrosation as a the main mechanism for inhibition of caspase-1 activity by NO-aspirin and our results demonstrated NO formation in monocytes incubated with this compound, it can not be excluded that a reaction product with NO^+ activity (N_2O_3, the reaction product of NO^+O_2, and even peroxynitrite formed from $NO^+O_2^-$) carry out this chemistry. The failure of DTT treatment to fully recover all YVADase activity, however, raises the possibility that NO may also suppress caspase-1 activation [14,15]. Indeed since caspase-1 activation is partially due to the autocalytic cleavage of the inactive proenzyme and pro-caspase-1 zymogens are themselves substrate for ICE [1-5], it can not be excluded that S-nitrosation/inhibition of activated ICE will reduces the amount of active enzyme that is further generated through this pathway [3]. Supporting this concept NO-aspirin markedly reduced the amount of the p20 subunit released in the cytoplasm of monocytes incubated with LPS. Moreover, we demonstrated that exposure to NO-aspirin, but not to aspirin, not only inhibited the release of ICE-regulated cytokines, IL-1 and IL-18, but more impressively resulted in an extensive downregulation of a wide array of pro-inflammatory chemokines and cytokines. However, since IL-8, IL-12, IFNγ and TNFα do not require an ICE-like peptidases for their maturation, it is likely that the inhibitory effect exerted by NO-aspirin was due to the inhibition of IL-1β and/or IL-18 production [7-11]. In this context the ability of NO-aspirin to suppress IL-1β, IL-8, IL-12, IL-18, and IFNγ production suggests that the NO-aspirin derivative may be effective in treating Th1-sustained diseases.

Conclusion

To summarize, in contrast to aspirin, NO-aspirin, that spares the gastric mucosa [10], causing the S-nitrosylation/inhibition of caspase-1, reduces the release of IL-1β and IL-18 in human monocyte/macrophage challenged with bacterial endotoxin . Since ICE activation is a limiting step in the process of maturation and secretion pro-inflammatory cytokines such as IL-1β and IL-18 , the caspase inhibition exerted by NO-aspirin may have important therapeutic implications for treatment of inflammatory disorders.

References

1. Dinarello CA. Biological basis for interleukin-1 in disease. Blood 1996;87:2095.
2. Thornberry NA, Bull HG, Calaycay JR, et al. A novel heterodimeric cysteine protease is required for interleukin-1 beta processing in monocytes. Nature 1992;356:768.
3. Salvesen GS, Dixit VM. Caspases: intracellular signaling by proteolysis. Cell 1997;91: 443.
4. Thornberry NA, Lazebnik Y. Caspases: enemies within. Science 1998;281:1312..
5. Alnemri ES, Livingston DJ, Nicholson DW,et al. Human ICE/CED-3 protease nomenclature. Cell 1997;87:171.
6. Dinarello CA, Margolis NH. Stopping the cuts: the recently discovered enzymes that process the precursors of inflammatory cytokines are good targets for the design of new anti-inflammatory therapeutic agents. Curr Biol 1995;5:587.
7. Elliott SN, McKnight W, Cirino G., Wallace JL. A nitric oxide-releasing nonsteroidal anti-inflammatory drug accelerates gastric ulcer healing in rats. Gastroenterology 1995;109: 524.
8. Wallace JL, McKnight W, Del Soldato P, Baydoun AR, Cirino G. 1995. Anti-thrombotic effects of a nitric oxide-releasing, gastric-sparing aspirin derivative. J Clin Invest 1995;96: 2711.
9. Fiorucci S, Santucci L, Federici B, et al. Nitric oxide (NO)-releasing NSAIDs inhibit interleukin-1β converting enzyme (ICE)-like cysteine proteases and protect endothelial cells from apoptotosis induced by TNFα. Aliment Pharmacol Ther 1999;13:421.
10. Fiorucci S, Antonelli E, Santucci L, et al. Gastrointestinal safety of NO-derived aspirin is related to inhibition of ICE-like cysteine proteases. Gastroenterology 1999;116:1089.
11. Fiorucci S, Santucci L, Antonelli E, et al. 2000. NO-aspirin protects from T-cell mediated liver injury by inhibiting caspase-dependent processing of Th1-like cytokines. Gastroenterology 2000;118:404.
12. Nicholson DW. From bench to clinic with apoptosis-based therapeutic agents. Nature 2000;407:810-816.
13. Fiorucci S, Santucci L, Cirino G, et al. IL-1beta converting enzyme is a target for nitric oxide-releasing aspirin: new insights in the antiinflammatory mechanism of nitric oxide-releasing nonsteroidal antiinflammatory drugs. J Immunol 2000 Nov 1;165(9):5245-54.
14. Dimmeler S, Haendeler J, Nehls M, Andreas M. Suppression of apoptosis by nitric oxide

via inhibition of interleukin-1 converting enzyme (ICE)-like and cysteine protease protein (CPP)-32-like proteases. J Exp Med 1997;185:601.

15. Kim Y-M., Talaninian RV, Li J, Billiar TR. Nitric oxide prevents IL-1β and IFNγ-inducing factor (IL-18) release from macrophages by inhibiting caspase-1 (IL-1β converting enzyme). J Immunol 1998;161:4122.
16. Young-Myeong K, Talanian RV, Billiar TR. Nitric oxide inhibits apoptosis by preventing increases in caspase-3-like activity via two distinct mechanisms. J Biol Chem 1997;272: 31138.
17. Rossig L, Fichtlscherer B, Breitschopf K, et al. Nitric oxide inhibits caspase 3 by S-nitrosation in vivo. J Biol Chem 1999;274:6823.
18. Dinarello CA, Novick D, Puren AJ, et al. Overview of interleukin-18: more than an interferon-γ inducing factor. J Leukoc Biol 1998;63:658.
19. Okamura H, Nagata K, Komatsu T, et al. A novel costimulatory factor for gamma interferon induction found in the livers of mice causes endotoxic shock. Infect Immun 1995;63:3966.
20. Puren AJ, Fantuzzi G, Gu Y, Su MS-S, Dinarello CA. Interleukin-18 (IFN gamma-inducing factor) induces IL-8 and IL-1beta via TNF alpha production from non-CD14$^+$ human blood mononuclear cells. J Clin Inv 1988;101:711.
21. Ghayur T, Banerjee S, Hugunin M, et al. Caspase-1 processes IFN-γ-inducing factor and regulates LPS-induced IFN-γ production. Nature 1997;386:619.
22. Gu Y, Kuida K, Tsutsui H, et al. Activation of interferon-gamma inducing factor is mediated by interleukin-1beta converting enzyme. Science 1997;275:206.
23. Li P, Allen H, Banerjee S, Franklin S, et al. Mice deficient in interleukin-1 converting enzyme (ICE) are defective in production of mature interleukin-1beta and resistant to endotoxic shock. Cell 1995;80:401.
24. Moncada S, Higgs EA. The L-arginine-nitric oxide pathway. N Eng J Med 1993;329:2002.
25. Rossig L, Fichtlscherer B, Breitschopf K, et al. Nitric oxide inhibits caspase 3 by S-nitrosation in vivo. J Biol Chem 1999;
26. Kolb H, Kolb-Bachofen V. Nitric oxide in autoimmune disease: cytotoxic or regulatory meditor? Immunol Today 1998;19:556.
27. Margolin N, Raybuck SA, Wilson KP, et al. Substrate and inhibitor specificity of interleukin-1-converting enzyme and related caspases. J Biol Chem 1997;272:7223.

POTENT INHIBITORY EFECTS OF A NO-NSAID (FLURBIPROFEN NITROXYBUTYLESTER) ON BONE RESORPTION *IN VITRO* AND *IN VIVO*

S.H. Ralston, K.J. Armour, R.J. van 't Hof, K.E. Armour, and P. Del Soldato

Introduction

Nonsteroidal anti-inflammatory drugs (NSAIDs) are widely employed in the treatment of musculoskeletal disorders but their use is limited by adverse effects on the gastrointestinal tract. A new class of NSAIDs, containing a nitric oxide (NO)-donor moiety (NO-NSAID) has been developed which retain the desired anti-inflammatory and analgaesic properties of NSAIDs but which do not cause gastrointestinal side effects, because the vasodilator effects of NO counteract the adverse effects of COX-1 inhibition on gastric mucosa [1,2]. Previous studies have shown that prostaglandins and nitric oxide play an important role in regulating bone cell function [3,4]. Since NO-NSAIDs combine the properties of a NO donor with those of a COX inhibitor, we have investigated the effects of the NO-NSAID flurbiprofen nitroxybutylester (HCT1026) on bone resorption *in vitro* and *in vivo*.

Materials and Methods

Osteoclast formation and activity was studied *in vitro* using the osteoblast-osteoclast co-culture as previously described [5] and *in vivo* studies were carried out on eight-week-old female C57/BL6 mice subjected to bilateral ovariectomy or sham ovariectomy. Three days after ovariectomy, treatment was started with equimolar quantities of HCT1026 (12.5 mg/kg body weight) or flurbiprofen (7.5 mg/kg body weight) given by intra-peritoneal injections in corn oil and were continued daily until termination of the experiment on day 21. Measurements of bone mineral content (BMC) and bone mineral density (BMD) at the left proximal tibial metaphysis was determined in anaesthetized mice by peripheral quantitative computed tomography (pQCT) using an XCT Research M bone densitometer with a voxel size of 100 μm and analysis software version 5.1.4. (Stratec Medizintechnik, Pforzheim, Germany). HCT1026 (flurbiprofen nitroxybutyl ester) and flurbiprofen were obtained from NiCox (Sophia Antipolis, France) and recombinant murine IL-1β and S-nitrosoacetyl penicillamine (SNAP) was obtained from Calbiochem-Novabiochem (Nottingham, U.K.). Cell culture media was from Gibco/BRL. Other reagents were purchased from Sigma (Poole, Dorset, UK). Statistical analyses were performed using Minitab version 12. Significant differences between groups were determined by ANOVA

B. Samuelsson et al. (eds.), Advances in Prostaglandin and Leukotriene Research, 179–182.

followed by post-hoc testing using Fisher's pair-wise comparisons. All data are presented as means ± SEM unless stated otherwise. Values of p less than 0.05 were considered significant.

Results

EFFECTS OF HCT1026 AND FLURBIPROFEN ON OSTEOCLAST FORMATION AND *IN VITRO*

Basal- and IL-1-induced osteoclast formation was significantly inhibited by HCT1026 at a concentrations of between 25-100μM (Table 1) and similar inhibitory effects were seen on bone resorption (data not shown). In contrast, flurbiprofen had no significant inhibitory effect on osteoclast formation under basal conditions and was significantly less effective than HCT1026 at inhibiting IL-1-induced osteoclast formation and bone resorption. Further studies (not shown) showed that the inhibitory effects of HCT1026 on osteoclast formation could not be mimicked by the combination of flurbiprofen and the pharmacological NO donor S-nitrosoacetyl penicillamine.

EFFECTS OF HCT1026 ON BONE *IN VIVO*

In order to determine if the inhibitory effects of HCT1026 on osteoclast activity *in vitro* were accompanied by inhibitory effects on bone loss *in vivo* we studied the effects of HCT1026 and flurbiprofen on ovariectomy-induced bone loss in mice. As expected, BMD and BMC fell significantly after ovariectomy when compared with sham operated animals. Administration of HCT1026 inhibited the reduction in BMD after ovariectomy whereas flurbiprofen had no significant effect (Figure 1).

Discussion

This study has shown that HCT1026 is a potent inhibitor of osteoclast formation and bone resorption *in vitro* and an inhibitor of ovariectomy induced bone loss *in vivo*. Previous studies have established that HCT1026 and flurbiprofen have similar potency with regard to inhibition of prostaglandin synthesis [1], yet in this study we found that the inhibitory effects of HCT1026 on osteoclastic formation and activity were much greater than those of flurbiprofen.

Parallel studies showed that HCT1026 was effective in preventing ovariectomy-induced bone loss *in vivo*. This also did not appear to be due to prostaglandin inhibition since equimolar doses of flurbiprofen had no significant protective effect on ovariectomy-induced bone loss. These data are in keeping with the results of previous workers who showed that IL-1-induced bone resorption could not be inhibited by Indomethacin [6,7]. Although high levels of NO have been shown to inhibit bone resorption *in vitro* [8], we found that the inhibitory effects of HCT1026 on osteoclast formation could not be reproduced by the combination of flurbirpofen and the NO-donor SNAP.

In summary, we have found that HCT1026 – a novel NO-NSAID – has potent inhibitory effects on osteoclast formation and bone resorption *in vitro* and prevents ovariectomy-induced bone loss in vivo. The effects of this compound cannot be reproduced by the combination of an NSAID and NO donor given separately implying that HCT1026 acts by a novel mechanism of action. Our data suggest that HCT1026 may therefore be of value in the prevention and treatment of disorders associated with osteoclast activation and increased bone resorption.

Table 1. Effects of HCT1026 and Flurbiprofen on Osteoclast Formation *In Vitro*

	Vehicle	HCT1026 25μM	HCT1026 100μM	Flurbiprofen 100μM
Basal	76 ± 15	54 ± 10*	39 ± 10 **	73 ± 23 *
IL-1	215 ± 68	125 ± 34**	32 ± 21**	123 ± 55 **

Values are mean ± SD osteoclast numbers. * $p < 0.05$, ** $p < 0.01$ from vehicle

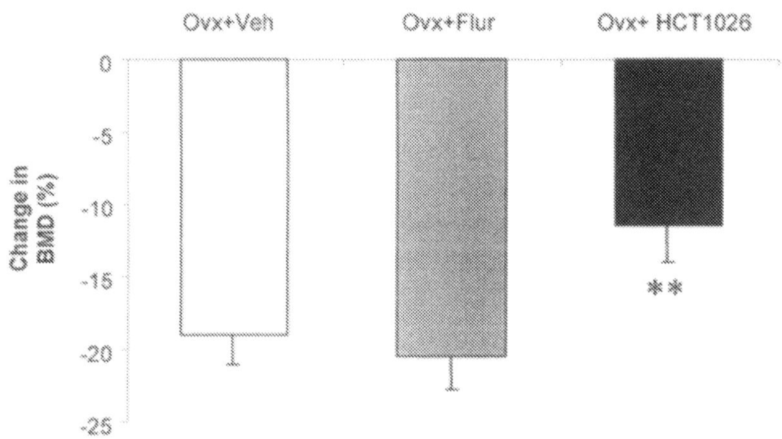

Figure 1. Effect of HCT1026 and flurbiprofen on bone loss *in vivo*. Values are mean ± SEM change in BMD, when compared with mice undergoing sham ovariectomy (6 animals per group). ** $p < 0.01$ from Ovx + veh. Ovx = ovariectomy; Flur = flurbiprofen 7.5mg/kg/day; HCT1026 12.5 mg/kg/day.

References

1. Wallace JL, Reuter B, Cicala C, et al. Novel nonsteroidal anti-inflammatory drug derivatives with markedly reduced ulcerogenic properties in the rat. Gastroenterology 1994 July;107(1):173-79.
2. Orr FW, Varani J, Gondek MD, et al. Chemotactic response of tumor cells to products of resorbing bone. Science 1979;203:176-79.

3. Raisz LG, Martin TJ. Prostaglandins in bone and mineral metabolism. In: Peck WA, editor. Bone and Mineral Research Annual. Amsterdam: Exerpta Medica, 1984: 286-310.
4. Ralston SH. Nitric oxide and bone: what a gas! Br J Rheumatol 1997;36:831-38.
5. van't Hof RJ, Ralston SH. Cytokine induced nitric oxide inhibits bone resorption by inducing apoptosis of osteoclast progenitors and suppressing osteoclast activity. J Bone Miner Res 1997;12:1797-804.
6. Gowen M, Wood DD, Ihrie EJ, et al. An interleukin-1 like factor stimulates bone resorption in vitro. Nature 1983;306:378-80.
7. Boyce BF, Aufdemorte TB, Garrett IR, et al. Effects of interleukin-1 on bone turnover in normal mice. Endocrinology 1989;125:1142-50.
8. MacIntyre I, Zaidi M, Towhidul Alam ASM, et al. Osteoclast inhibition: an action of nitric oxide not mediated by cyclic GMP. Proc Natl Acad Sci USA 1991;88:2936-40.

Lipid Peroxidation and Cyclooxygenase-2 Activity in Human Hepatocellular Carcinoma Cell Lines With and Without Multiple Drug Resistance: Correlation With Cell Growth

Emanuela Masini, Ornella Fantappiè, Iacopo Sardi, Laura Raimondi, Valentina Fabbroni, Pier Francesco Mannaioni, and Roberto Mazzanti

Introduction

Intrinsic or acquired resistance to antineoplastic drugs is the main cause of failure in cancer chemotherapy. Human hepatocellular carcinoma (HCC) has been increasing on a worldwide scale and it is among the most chemoresistant tumors. One of the main mechanisms of drug resistance is related to the over-expression of the gene that encodes a transmembrane *P*-glycoprotein which is involved in drug extrusion from the cells that possess the multiple drug-resistant (MDR1) phenotype [1]. Cancers with an MDR phenotype are more resistant to several anticancer drugs. Free radical formation is purported to be involved in cytotoxic effects of anticancer agents. However in HCC cells the expression of MDR1 phenotype *per se* does not increase significantly the resistance to iron-stimulated lipid peroxidation [2]. Moreover hepatitis B and C viruses are found to be important factors of hepatocarcinogenesis, through the induction of cell proliferation and angiogenesis. Cyclooxygenase-2 (COX-2) has now been identified as expressed in a number of cancers, including HCC [3], but whether COX-2 activity affects cell growth and susceptibility to free radical attack in human HCC cell lines has not been elucidated. To assess these points, an HCC MDR1 cell line highly resistant to doxorubicin, P1(0.5), and the parental drug-sensitive cells, P5, were studied for their susceptibility to lipid peroxidation and for the effect afforded by nimesulide, a COX-2 inhibitor.

Methods

Experiments were performed on a HCC cell line (PLC/PRF/5). The P1(0.5) MDR clone, selected and maintained in 0.5 µg/ml doxorubicin [2] and the parental drug-sensitive cells (P5) were used. Cell lines were cultured in DMEM supplemented with 10% fetal calf serum (FCS), 100 U of penicillin, 10 µg/ml streptomycin, and 0.25 µg/ml amphotericin B, at 37 °C in 5% CO_2. Susceptibility to the free radical attack was assessed by measuring lipid peroxidation by the determination of thiobarbituric acid-reactive substances (TBARS) in the reaction mixture after 60 minutes of incubation at 37 °C of cells exposed to ascorbate-Fe

B. Samuelsson et al. (eds.), Advances in Prostaglandin and Leukotriene Research, 183–186.

(0.5 mM and 25 μM $FeSO_4$) [4]. Alphatocopherol (vitamin E) content in the cells was measured using a HPLC method [2]. PgE_2 production was determined in the supernatants by a RIA method. P5 and P1(0.5) cell growth was evaluated by [^{3}H]thymidine incorporation assay. Briefly, 10^4 cells were seeded in 24-well plates and left to grow to subconfluence. After 24-hour incubation in a steady-state medium (0.1% FCS), the medium was substituted with DMEM containing 10% FCS, added with epidermal growth factor (EGF, 100 nM) in the presence or absence of nimesulide, a COX-2 inhibitor, at the noted concentrations. At the end of the incubations, the cells were pulsed for 4 hours with 0.5 μCi per well of [*methyl*-^{3}H]thymidine (specific activity, 46 Ci/mmol). The medium was then removed, and DNA was precipitated with 3% trichloroacetic acid (TCA) and extracted with 1 ml of 0.3 M NaOH. The recovered radioactivity was measured in a beta scintillation counter. The experiments were performed in quadruplicate, and the values were expressed as dpm/well (mean+SE). Statistical analysis was performed using a multifactor ANOVA (Sheffe's *t* test) and Student's *t* test for unpaired values. $P < 0.05$ was considered significant.

Results and Discussion

Concentrations of TBARS measured at time 0 was similar in P5 and P1(0.5) cells. Exposure to ascorbate Fe of P5 cells significantly increased TBARS production, which was left unchanged in ascorbate-Fe treated P1(0.5) cells (Figure 1a). Vitamin E cell content was significantly higher ($p < 0.05$) in P1(0.5) cells expressing the MDR phenotype (Figure 1b).

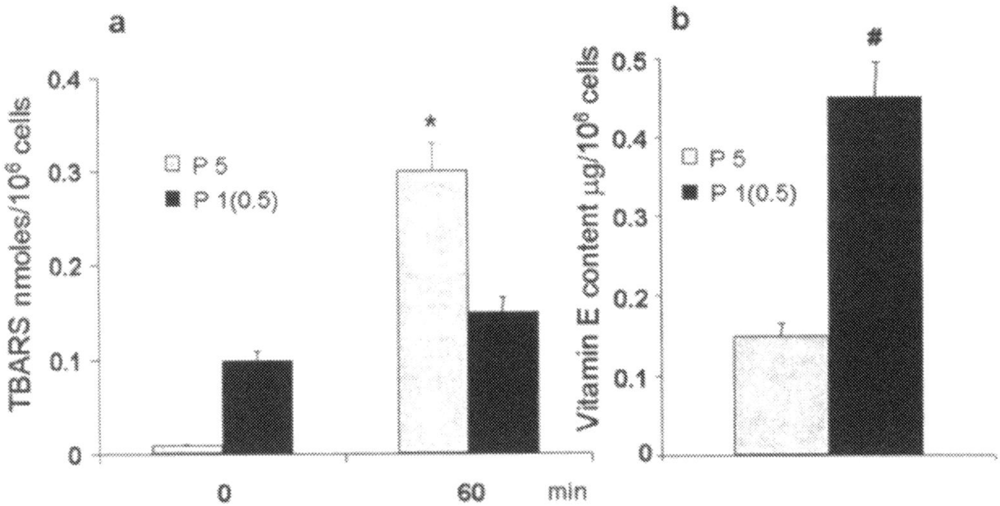

Figure 1. TBARS concentration at time 0 and after 60 minutes exposure to ascorbate-Fe (panel a), and vitamin E content (panel b). * $p < 0.05$ versus 0 minutes.; # $p < 0.05$ versus P5.

Basal PgE_2 production was higher in P1(0.5) cells than in P5 cells. The exposure to 100 nM EGF increased PgE_2 production in both cell clones but the amounts of PgE_2 produced were significantly higher in P1(0.5) cells. The treatment with nimesulide (10 uM), a COX-2 inhibitor, blunted the EGF-induced PgE_2 production in both cell clones (data not shown). The involvement of COX-2 in cell growth is shown by the effect of nimesulide on [^{3}H]thymidine incorporation induced by EGF: nimesulide significantly reduced EGF-stimulated thymidine incorporation in P1(0.5) cell clone but not in P5 one (Figure 2).

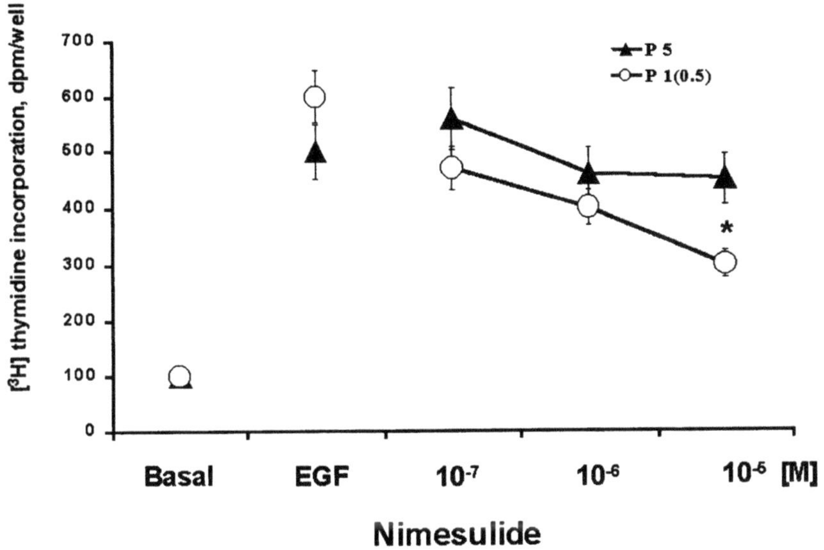

Figure 2. Effect of nimesulide on EGF(100 ng/ml)-stimulated [^{3}H]thymidine incorporation in P5 and P1(0.5) cell clones. * $p < 0.01$.

These data suggest that the resistance of HCC to undergo lipid peroxidation is related to an increased vitamin E content rather than to MDR1 phenotype; while the MDR1 phenotype is associated with a COX-2 up-regulation and COX-2 control of mitogen-activated cell proliferation. In conclusion COX-2 inhibitors could have a role in the therapy of MDR HCC, by increasing the responsiveness of the cells to chemotherapy.

References

1. Mazzanti R, Gatmaitan AZ, Croop JM, Shu H, Arias IM. Quantitative image analysis in Rhodamine 123 transport by adriamicin-sensitive and resistant NIH3T3 and human hepatocellular carcinoma (Alexander) cells. J Cell Pharmacol 1974;1:50-56.
2. Mazzanti R, Fantappiè O, Fabrizio P, Relli P. Susceptibility to lipid peroxidation of human hepatocellular carcinoma cell lines with different levels of multiple drug-resistant phenotype. Lab Invest 1995;73:419-23.
3. Shiota G, Okubo M, Noumi T, et al. Cyclooxygenase-2 expression in hepatocellular

carcinoma. Hepatogastrenterology 1999;46:407-12.

4. Ohkawa H, Ohishi N, Yagi K. Assay for lipid peroxides in animal tissues by thiobarbituric acid reaction. Anal Biochem 1979;95:351-58.

PULMONARY PROSTACYCLIN SYNTHASE OVEREXPRESSION IMPROVES SURVIVAL FOLLOWING BUTYLATED HYDROXYTOLUENE EXPOSURE

Robert L. Keith, York E. Miller, Tracey L. Gesell, Mark D. Moore, Alvin M. Malkinson, and Mark W. Geraci

Lung cancer is the leading cause of cancer death in men and women in North America [1]. Although primary prevention of tobacco smoking and smoking cessation are the most potent interventions available, most lung cancers are currently diagnosed in former smokers [1], underscoring the need for effective chemoprevention strategies. Mouse models of lung carcinogenesis display both histologic and molecular genetic similarities to adenocarcinoma [2], the most common histologic type of human lung cancer. In these models, modulation of inflammation with either nonselective COX-1 and COX-2 or selective COX-2 inhibition, cause reductions in lung tumor multiplicity [3,4]. To date, large-scale interventional trials of COX inhibition in human lung cancer chemoprevention have not been performed.

Prostaglandin H_2 (PGH_2), the product of the COX enzymes, is metabolized to multiple eicosanoids, some of which may prove critical in tumorigenesis. Previous studies demonstrate that prostacyclin (PGI_2), a PGH_2 metabolite, has properties relevant to carcinogenesis, including suppression of inflammation [5], platelet inhibition [6], metastasis prevention [7], and growth inhibition of established micrometastases [8]. To investigate the role of PGI_2 in murine lung carcinogenesis, we produced transgenic mice expressing rat prostacyclin synthase (PGIS) under control of the human surfactant apoprotein C (SPC) promoter [9]. PGIS, a 52 kDa, membrane-associated P450-like enzyme, is the final committed enzymatic step in the production of PGI_2. The human SPC promoter directs expression of transgenes to alveolar type II and Clara cells [10], which are the progenitors for human and mouse lung adenocarcinomas. PGI_2 and PGE_2 production in normal lung and human non-small cell lung cancers (NSCLC) has been examined. In normal lung, we found that PGI_2 is the major prostaglandin present (3450 ng/g lung tissue), with very low levels of PGE_2 being observed (430 ng/g). In contrast, NSCLC cell lines produce very low amounts of PGI_2 [11]. In cultured NSCLC cell lines, we found a shift in the prostaglandin profile, with very low levels of PGI_2 (333 ng/g) and high levels of PGE_2 (11,952 ng/g). One explanation for the high PGE_2 is constituitively high expression of cytosolic phospholipase A_2 and COX-2 due to Ki-*ras* mutations [12].

Transgenic mice, with a 50% increase in lung PGIS activity as exhibited by a 1.5-fold increase in 6-keto $PGF_{1\alpha}$ (the stable metabolite of PGI_2) compared to wildtype littermates (Figure 1), were exposed to an inflammatory insult. Tg^+ and Tg^- littermates, 8

B. Samuelsson et al. (eds.), Advances in Prostaglandin and Leukotriene Research, 187–190.

to 12 weeks of age, were subjected to an intraperitoneal injection of butylated hydroxytoluene (BHT) (150-200 mg/kg mouse weight). BHT induces reversible pulmonary damage characterized by alveolar type I cell necrosis, selective pulmonary inflammation, and hyperplasia of alveolar type II cells [13]. Controls consisted of mice injected with vehicle alone. Experimental mice had bronchoalveolar lavage performed six days after BHT treatment. Cell counts (with differentials) and protein concentrations were obtained on the lavage fluid.

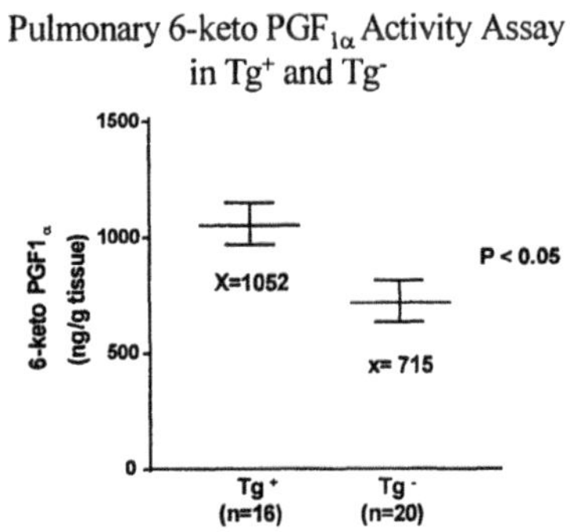

Figure 1. Prostacyclin synthase activity in transgenic and wildtype littermates illustrating a significant increase in the transgenic animals ($p<0.05$).

Tg^+ animals exhibited a distinct survival advantage compared to Tg^- animals. When exposed to BHT at a dose of 200mg/kg, all of the Tg^+ survived (12/12) and all of the Tg^- animals died (8/8). When the dose of BHT was lowered to 150mg/kg, all the Tg^+ (12/12) survived, as did the majority (5/8) of the Tg^-. Bronchoalveolar lavage of both groups of experimental animals six days after BHT at a dose of 150 mg/kg revealed significantly elevated cell counts in the transgenics (14.2×10^8 versus 7.62×10^8, $p < 0.05$) with > 95% macrophages on differential (Figure 2). The significant differences in cell counts were not evident three days after the BHT injection and there were no significant differences in the differentials at any of the time points. Higher cell counts were associated with higher protein levels in all animals studied. Whether improved survival in the transgenics is due to alterations in macrophage phenotype or to induction of resistance to lung parenchymal injury is not known.

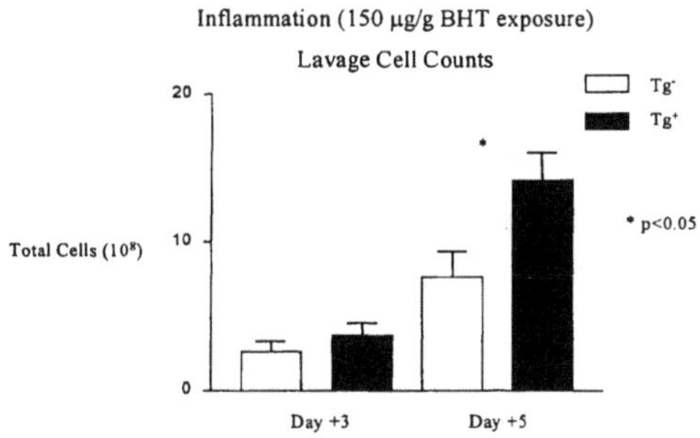

Figure 2: Cell counts from bronchoalveolar lavages performed three and five days following BHT exposure. On day +5, the transgenic animals had significantly higher cell counts ($p < 0.05$).

While COX inhibition is a proven anti-inflammatory strategy, our findings suggest that manipulation of the arachidonic acid pathway downstream of COX by selectively increasing pulmonary prostacyclin levels can improve survival after an inflammatory insult. In addition, this may have effects on other lung disorders in which inflammation is thought to play a role, including development of emphysema, bronchitis, asthma, and lung cancer.

References

1. Landis SH, et al. Cancer facts & figures - 1999. CA Cancer J Clin 1999;49:1-36.
2. Malkinson AM. Molecular comparison of human and mouse pulmonary adenocarcinomas. Exp Lung Res 1998;24:541-55.
3. Rioux N, Castonguay A. Prevention of NNK-induced lung tumorigenesis in A/J mice by acetylsalicylic acid. Cancer Res 1998;58:5354-60.
4. Duperron C, Castonguay A. Chemoprotective efficacies of aspirin and sulindac against lung tumorigenesis in A/J mice. Carcinogenesis 1997;18:1001-6.
5. Vane JR. Prostacyclin: a hormone with a therapeutic potential. J Endocrin 1982;95:3P-43P.
6. Dusting GJ. et al. Prostacyclin: its biosynthesis, actions, and clinical potential. Adv Prostaglandin Thromboxane Leukot Res 1982;10:59-106.
7. Honn KV, Cicone B, Skoff A. Prostacyclin: a potent antimetastatic agent. Science 1981; 212:1270-72.

8. Schirner M, Schneider MR. Inhibition of metastasis by cicaprost in rats with established SMTA2 mammary carcinoma growth. Cancer Detect Prev 1997;21:44-50.
9. Geraci MW, et al. Selective pulmonary prostacyclin synthase overexpression in transgenic mice protects against development of hypoxic pulmonary hypertension. J Clin Invest 1999; 103:1509-15.
10. Korfhagen TR, et al. cis-acting sequences from a human surfactant protein gene confer pulmonary-specific gene expression in transgenic mice. Proc Natl Acad Sci USA 1990;87: 6122-26.
11. Hubbard WC, et al. Evidence for prostanoid biosynthesis as a biochemical feature of certain subclasses of non-small cell carcinomas of the lung as determined in established cell lines derived from human lung tumors. Cancer Res 1989;49:826-32.
12. Heasley LE, et al. Induction of cytosolic phospholipase A_2 by oncogenic ras in human non-small cell lung cancer. J Biol Chem 1997;272:14501-4.
13. Marino AA, Mitchell JT. Lung damage in mice following intraperitoneal injection of butylated hydroxytoluene. Proc Soc Exp Biol Med 1972;140:122-25.

REACTIVE PRODUCTS OF THE ISOPROSTANE PATHWAY: ISOKETALS AND CYCLOPENTENONE A_2/J_2-ISOPROSTANES

L. Jackson Roberts, II, Yan Chen, Olivier Boutaud, Sean S. Davies, Jason D. Morrow, John A. Oates, and Cynthia Brame

Introduction

Isoprostanes (IsoPs) are prostaglandin-like compounds that are formed non-enzymatically by free radical catalyzed peroxidation of arachidonic acid. The archetypical compounds discovered to be formed by this pathway were prostaglandin F_2-like compounds, F_2-IsoPs [1]. An important facet of the formation of IsoPs is that, unlike prostaglandins, they are initially formed *in situ* esterified to phospholipids and then released preformed [2]. Key intermediates in the formation of IsoPs, analogous to the formation of prostaglandins via the cyclooxygenase pathway, are bicyclic endoperoxides PGH_2-like compounds (H_2-IsoPs). H_2-IsoPs are reduced to form F_2-IsoPs. However, we have found that this reduction is not completely efficient such that the H_2-IsoPs undergo rearrangement *in vivo* to form prostaglandin E_2-like and D_2-like compounds (E_2/D_2-IsoPs) and thromboxane-like compounds (isothromboxanes) [3,4].

Formation of A_2/J_2-Isoprostanes *in Vivo*

The cyclopentenone (CP) prostaglandins PGA_2 and PGJ_2 are formed by dehydration of PGE_2 and PGD_2, respectively. Interest in these compounds stems from the fact that they have been found to exert interesting biological actions such as modulation of tumor cell proliferation and activation of PPARγ [5,6]. Although the formation of PGA_2 and PGJ_2 can be demonstrated to occur *in vitro*, whether they are actually formed *in vivo* has been the subject of debate for nearly three decades [7-9]. A single report convincingly demonstrated the occurrence of Δ^{12}-PGJ_2 in human urine [10]. However, it cannot be concluded from this finding that Δ^{12}-PGJ_2 was actually formed systemically because it may have arisen from dehydration of PGD_2 in bladder urine. This possibility is suggested by the finding that PGD-synthase, which catalyzes the dehydration of PGD_2, is present in high concentrations in human semen and levels of Δ^{12}-PGJ_2 in urine were found to be substantially higher in males than in females [11].

We initially established that CP-IsoPs were generated during oxidation of arachidonic acid *in vitro* utilizing a number of analytical approaches including electron

B. Samuelsson et al. (eds.), Advances in Prostaglandin and Leukotriene Research, 191–195.

impact mass spectrometery [12]. We then explored whether A_2/J_2-IsoPs are also formed *in vitro* and *in vivo* after treatment of rats with CCl_4, a well-established model of free radical induced injury to the liver. Analysis of CP-IsoPs and E_2/D_2-IsoPs was performed following TLC purification by GC/MS using [$^2H_4$] PGA_2 and [$^2H_4$] PGE_2 as internal standards, respectively. The results obtained are shown in Figure 1. CP-IsoPs and E_2/D_2-IsoPs were present in detectable quantities esterified in livers from control animals and levels of both increased dramatically and to a similar extent following administration of CCl_4.

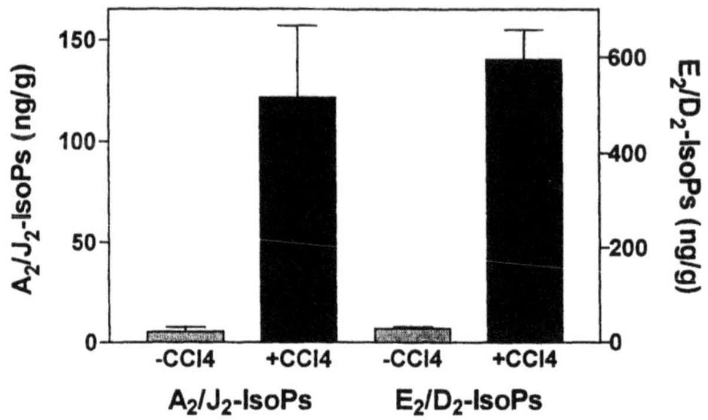

Figure 1. Levels of A_2/J_2-IsoPs and E_2/J_2-IsoPs esterified in the liver of control rats and rats treated with CCl_4.

We then examined the time-course of conjugation of one the A_2-IsoPs (15-A_{2t}-IsoP, 8-iso-PGA_2) (Figure 2A) with glutathione in the presence of gluathione *S*-transferase of and its rate of adduction to albumin (Figure 2B). In the absence of glutathione *S*-transferase, the rate of Michael addition with glutathione was neglible over a 30-minute time period.

Whereas high levels of E_2/D_2-IsoPs are present in the circulation in rats following treatment with CCl_4 [3], CP-IsoPs could not be detected even under this situation of severe oxidant injury. We hypothesized that this may be attributed to rapid formation of polar conjugates of CP-IsoPs, e.g. with glutathione, following their hydrolysis from phospholipids. To test this hypothesis, we infused a tracer amount of radiolabelled 15-A_{2t}-IsoP into a normal human volunteer and determined the amount of radioactivity excreted into the urine that would not extract into methylene chloride. We found that essentially all (95%) of the radioactivity excreted was not extractable, suggesting a predominance of polar conjugates. This finding may explain the failures in the past to provide convincing evidence for the formation of PGA_2 and PGJ_2 *in vivo* by attempts to detect these compounds in free form.

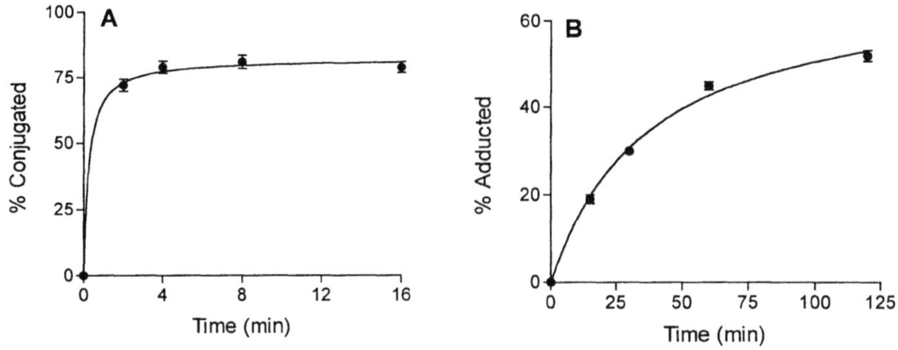

Figure 2. Time-course of conjugation of 15-A_{2t}-IsoP with glutathione in the presence of glutathione *S*-tranferase (A) and adduction to bovine serum albumin (B).

Formation of Isoketals *in Vitro* and *in Vivo*

Salomon and colleagues had previously described the formation of acyclic γ-ketoaldehdyes *in vitro* as decomposition products PGH_2 [13]. Therefore, we explored whether similar compounds, which we previously termed isolevuglandins but now refer to as isoketals (IsoK), are formed as decomposition products of H_2-IsoPs (Figure 3).

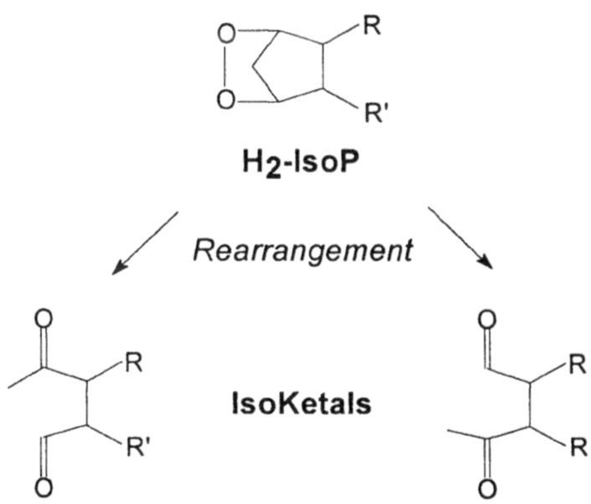

Figure 3. Formation of Isoketals as decomposition products of H_2-IsoPs

Our interest in IsoKs relates to the fact that these γ-ketoaldehydes are extremely reactive molecules which rapidly adduct to lysine residues on proteins, which may have relevance to the pathobiology of oxidant injury. Using analyses based on GC/MS, we were able to demonstrate the formation of abundant quantities of IsoKs *in vitro* during oxidation of arachidonic acid *in vitro* [14]. The amounts formed were intermediate between the amounts of F_2-IsoPs and E_2/D_2-IsoPs formed. However, exhaustive efforts to detect these compounds during oxidation of even simple biological systems *in vitro*, e.g. oxidation of liver microsomes and LDL, were not successful. We speculated that this may be due to rapid adduction of IsoKs to proteins. We tested this hypothesis by comparing the rate of adduction of IsoKs and 4-hydroxynonenal (HNE) to albumin *in vitro*. HNE is considered to be one of the most reactive products of lipid peroxidation that has been identified. However, HNE can be detected in free form in biological tissues and fluids. We found that a synthetic IsoK adducted to albumin within about 30 seconds whereas about 50% of HNE was still in free unadducted form after 80 minutes of incubation.

The above finding supported the notion that our failure to detect free IsoKs in free form in biological systems was due to rapid adduction to proteins. We therefore undertook to elucidate the nature of these protein adducts. We found that IsoKs initially form a reversible Schiff base adduct with lysine and then proceeds through a pyrrole, which undergoes facile oxidation to lactam and hydroxylactam adducts [14,15].

We then developed LC/MS/MS methodology for the detection and measurement of lysyl-IsoK Schiff lactam adducts following complete enzymatic digestion of proteins to individual amino acids using pronase and leucine aminopeptidase. Using this approach, lactam adducts were present in detectable quantities in both normal rat and human plasma at levels of 199 ± 78 pg/ml and 561 ± 101 pg/ml, respectively.

In summary, these studies have identified two new classes of reactive compounds that are formed *in vivo* as products of the IsoP pathway, namely CP-IsoPs (A_2/J_2-IsoPs) and IsoKs. This provides a basis to begin to explore the biological consequences of the formation of these novel molecules as it relates to the pathobiology of oxidant injury.

Acknowledgments

This work was supported by grants GM42056, GM15431, DK26657, and DK48831 from the National Institutes of Health.

References

1. Morrow JD, Hill KE, Burk RF, Nammour TM, Badr KF, Roberts LJ, II. A series of prostaglandin F_2-like compounds are produced *in vivo* in humans by a non-cyclooxygenase free radical catalyzed mechanism. Proc Natl Acad Sci USA, 1990;87:9383-87.
2. Morrow JD, Awad JA, Boss HJ, Blair IA, Roberts LJ, II. Non-cyclooxygenase derived prostanoids (F_2-isoprostanes) are formed *in situ* on phospholipids. Proc Natl Acad Sci USA 1992;89:10721-25.
3. Morrow JD, Minton TA, Mukundan CR, et al. Free radical induced generation of isoprostanes in vivo: Evidence for the formation of D-ring and E-ring isoprostanes. J Biol Chem 1994;269:4317-26.
4. Morrow JD, Awad JA, Wu A., Zackert WE, Daniel VC, Roberts LJ, II. Free radical-induced generation of thromboxane-like compounds (isothromboxanes) in vivo. J Biol

Chem 1996;271:23185-90.

5. Negishi M, Koizumi T, Ichikawa A. Biological actions of delta 12-prostaglandin J_2. J Lip Mediat Cell Signal 1995;12:443-48.
6. Forman BM., Tontonoz P, Chen J, Vrun RP, Spiegelman BM, Evans RM. 15-deoxy-$\Delta^{12,14}$-PGJ_2 is a ligand for the adipocyte determination factor PPARγ. Cell 1995;83:803-12.
7. Attallah A, Payakkapan W, Lee J, Carr A, Brazelton E. PGA: fact, not artifact. Prostaglandins 1974;5:69-71.
8. Middledtich BS. PGA: fact or artifact? Prostaglandins 1975;9:409-11.
9. Maxey KM, Hessler E, MacDonald J, Hitchingham L. The nature and composition of 15-deoxy-$\Delta^{12,14}$-PGJ_2. Prostaglandins Other Lip Med 2000;62:15-21.
10. Hirata Y, Hayashi H, Ito S, et al. Occurrence of 9-deoxy-$\Delta^{9,12}$-13,14dihydroprostaglandin D_2 in human urine. J Biol Chem 1988;263:16619-25.
11. Togkugawa Y, Kunsihige I, Kubota Y, et al. Lipcalin-type prostaglandin D synthase in human male reproductive organs and seminal plasma. Biol Reprod 1998;58:600-7.
12. Chen Y, Morrow JD, Roberts LJ, II. Formation of reactive cyclopentenone compounds *in vivo* as products of the isoprostane pathway. J Biol Chem 1999;274:10863-68.
13. Salomon RG, Miller RG, Zagorski MG, Coughlin DJ. Sovent induced fragmentation of prostaglandin endoperoxides. New aldehyde products from PGH_2 and a novel intramolecular 1,2-hydride shift during endoperoxide fragmentation in aqueous solution. J Am Chem Soc 1984;106:6049-60.
14. Brame CJ, Salomon RG, Morrow JD, Roberts LJ, II. Identification of extremely reactive γ-ketoaldehydes (isolevuglandins) as products of the isoprostane pathway and characterization of their lysyl protein adducts. J Biol Chem 1999;274:13139-46.
15. Boutand O, Brame CJ, Salomon RG, Roberts LJ, II, Oates JA. Characterization of lysyl adducts formed from prostaglandin H_2 via the levuglandin pathway. Biochemistry 1999; 38 9389-96.

INCREASED ISOPROSTANE 8-EPI-PGF$_{2A}$ IN STATIN-INDUCED MYOPATHY

Helmut Sinzinger, Graziana Lupattelli, Fahdi Chehne

Introduction

The most prevalent side effect of statins are muscle pains occurring in presence but also in absence [1] of elevated creatine phosphokinase (CK). In one patient statin-induced muscular pains improved significantly after vitamin E [2]. Thereafter, examining another patient we found for the first time a pronounced increase in 8-epi-PGF$_{2\alpha}$, disappearing after withdrawal of the respective statin. Earlier it has been reported that rhabdomyolysis is associated with an increase in isoprostane (IP) formation [3]. Therefore, we assessed the IP 8-epi-PGF$_{2\alpha}$ in plasma, serum, and urine after extraction and purification using a specific immunoassay.

Material and Methods

In 30 patients (16 males, 14 females; 34-58 years) with heterozygous familial hypercholesterolemia diagnosed receptor positive after radiolabeling of autologous LDL with ^{111}In (as described by us earlier) prevalues of IP before initiating statin (atorvastatin [A], fluvastatin [F], lovastatin [L], pravastatin [P], simvastatin [S]) therapy (-2, -1, 0 days) as well as after starting or the onset of symptoms or drug withdrawal at daily intervals up to 2 weeks and thereafter weekly up to 1 month were obtained. Blood was drawn in the morning after an overnight fasting and processed as described [4]. Values are given in pg/ml (plasma), pg/ml (serum), and pg/mg creatinine (urine).

Results

Our findings indicate that in almost half of the patients (n = 13) experiencing muscular side effects with or without CK-elevation there is a significant increase in 8-epi-PFG$_{2\alpha}$, which returns stepwise back to prevalues after stopping statin treatment. Withdrawing the drug resulted in a stepwise decrease in the respective IP-values, almost normalizing within 2, and completely normalizing within 4 weeks. CK (if elevated) rapidly normalized as well. This effect was seen with all the statins available at the moment, although patients showed the increase only with one or the other compound.

B. Samuelsson et al. (eds.), Advances in Prostaglandin and Leukotriene Research, 197–199.

Two patient examples (Figures 1 and 2) show the typical kinetics if 8-epi-$PGF_{2\alpha}$ increases upon statin therapy. Other patients (data not shown) do not respond at all (n = 17).

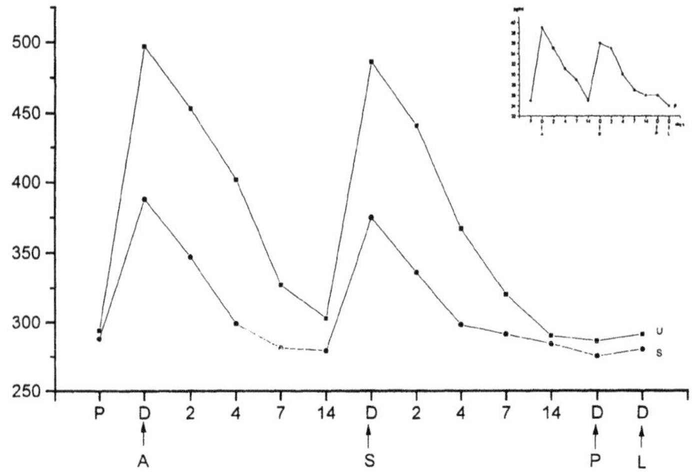

Figure 1. A and S, but not P and L are increasing 8-epi-$PGF_{2\alpha}$ in plasma (p) [insert], urine (u) and serum (s) (for units see Materials and Methods). Withdrawal of the respective statin results in fast normalization of IP in all the compartments examined. (Patient no. 7, male, 39 years old; isolated hypercholesterolemia)

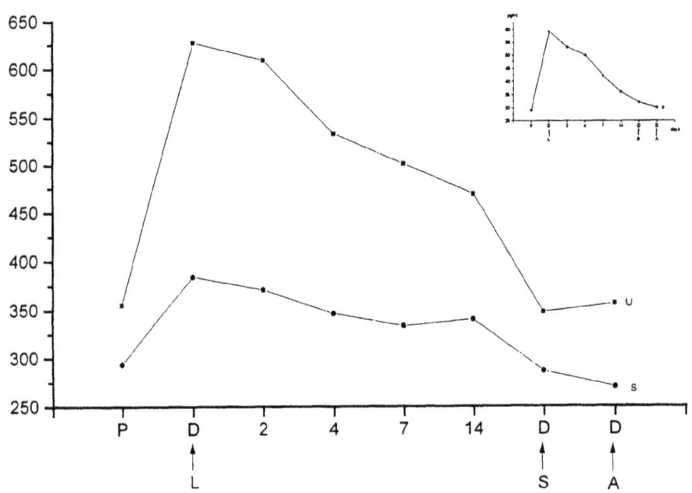

Figure 2. L but not S and A are causing an 8-epi-$PGF_{2\alpha}$ increase in this patients (for units see Materials and Methods). Patient female, 47 years old; isolated familial heterozygous hypercholesterolemia)

Discussion

Our findings for the first time show that during the symptomatic period of muscle pains with or without concomitant CK-elevation during statin therapy, 8-epi-$PGF_{2\alpha}$ is severely increased indicating a significant *in vivo* oxidation injury [5]. Statins among others inhibit several isoprenoid end-products, including ubiquinone, a potent scavenger of free radicals and inhibitor of lipid peroxidation. A decrease in plasma, but not in tissue, was found after use of various statins. A reversal of symptoms by mevalonate [6], ubiquinone [7], and vitamin E [1] has been reported. In our patients we did not determine ubiquinone. To assess oxidizability of lipoproteins in this setting is difficult, as the lipid-lowering effect, the drug itself as well as the lowering of ubiquinone, among others, may contribute. A screening investigation exhibited, that in (very few) statin treated patients without any side effect, isoprostane 8-epi-$PGF_{2\alpha}$ is elevated (unpublished data). Under normal circumstances lipoproteins (at least low density and high density lipoproteins) are becoming more resistant to oxidation injury during statin treatment and isoprostanes are decreasing. Apparently, oxidation injury is involved in statin-induced myopathy. Background and pathogenesis of this phenomenon urgently need to be elucidated.

Acknowledgments

The valuable help of Eva Unger in preparing and typing the manuscript and Bärbel John in computing the figures is gratefully acknowledged.

References

1. Sinzinger H. Does vitamin E beneficially affect muscle pains during HMG-Co-enzyme-A-reductase inhibitors without CK-elevation? Atherosclerosis 2000;149:225-26.
2. Sinzinger H, Lupattelli G, Chehne F. Increased lipid peroxidation in a patient with CK-elevation and muscle pain during statin therapy. Atherosclerosis 2000; in press.
3. Holt S, Reeder B, Wilson M, Harvey S, Morrow JD, Roberts LJ II, Moore K. Increased lipid peroxidation in patients with rhabdomyolysis. Lancet 1999;353:358-59.
4. Oguogho A, Mehrabi M, Sinzinger H. Increased plasma, serum and urinary 8-epi-prostaglandin $F_{2\alpha}$ in heterozygous hypercholesterolemia. Wiener klin Wochenschr 1999;111: 113-18.
5. Sinzinger H. Two different types of exercise-induced muscle pain without myopathy and CK-elevation during HMG-Co-enzyme-A-reductase inhibitor treatment. Atherosclerosis 1999;143:459-60.
6. Dujovne CA. New lipid lowering drugs and new effects of old drugs. Curr Opin Lipidol 1997;8:362-68.
7. Bliznakov E, Wilkins DJ. Biochemical and clinical consequences of inhibiting coenzyme Q_{10} biosynthesis by lipid-lowering HMG-CoA reductase inhibitors (statins): a critical overview. Adv Therapy 1998;15:218-28.

Subject Index

Medical Science Symposia Series

1. A.M. Gotto, C. Lenfant, R. Paoletti (eds.) and M. Soma (ass.ed.): *Multiple Risk Factors in Cardiovascular Disease.* 1992 ISBN 0-7923-1938-9
2. A.L. Catapano, A.M. Gotto, Jr., L.C. Smith and R. Paoletti (eds.): *Drugs Affecting Lipid Metabolism.* 1993 ISBN 0-7923-2232-0
3. T. Godfraind, S. Govoni, R. Paoletti and P.M. Vanhoutte (eds.): *Calcium Antagonists. Pharmacology and Clinical Research.* 1993 ISBN 0-7923-2259-2
4. D. Galmarini, L.R. Fassati, R. Paoletti and S. Sherlock (eds.): *Drugs and the Liver: High Risk Patients and Transplantation.* 1993 ISBN 0-7923-2307-6
5. P.M. Vanhoutte, P.R. Saxena, R. Paoletti, N. Brunello (eds.) and A.S. Jackson (ass.ed.): *Serotonin. From Cell Biology to Pharmacology and Therapeutics.* 1993 ISBN 0-7923-2518-4
6. A.G. Dalgleish, A. Albertini and R. Paoletti (eds.): *The Impact of Biotechnology on Autoimmunity.* 1994 ISBN 0-7923-2724-1
7. P.G. Crosignani, R. Paoletti, P.M. Sarrel, N.K. Wenger (eds.), M. Meschia and M. Soma (ass.eds.): *Women's Health in Menopause. Behaviour, Cancer, Cardiovascular Disease, Hormone Replacement Therapy.* 1994 ISBN 0-7923-3068-4
8. A.M. Gotto Jr., C. Lenfant, A.L. Catapano and R. Paoletti (eds.): *Multiple Risk Factors in Cardiovascular Disease.* 1995 ISBN 0-7923-3503-1
9. T. Godfraind, G. Mancia, M.P. Abbracchio, L. Aguilar-Bryan and S. Govoni (eds.): *Pharmacological Control of Calcium and Potassium Homeostasis. Biological, Therapeutical, and Clinical Aspects.* 1995 ISBN 0-7923-3604-6
10. A.M. Gotto Jr., R. Paoletti, L.C. Smith, A.L. Catapano and A.S. Jackson (eds.): *Drugs Affecting Lipid Metabolism. Risk Factors and Future Directions.* 1996 ISBN 0-7923-4167-8
11. R. Paoletti, P.G. Crosignani, P. Kenemans, G. Samsioe, M. Soma and A.S. Jackson (eds.): *Women's Health and Menopause.* Risk Reduction Strategies. 1997 ISBN 0-7923-4697-1
12. A.M. Gotto Jr., C. Lenfant, R. Paoletti, A.L. Catapano and A.S. Jackson (eds.): *Multiple Risk Factors in Cardiovascular Disease.* Strategies of Prevention of Coronary Heart Disease, Cardiac Failure. 1998 ISBN 0-7923-5023-5
13. R. Paoletti, P.G. Crosignani, P. Kenemans, N.K. Wenger and A.S. Jackson (eds.): *Women's Health and Menopause.* Risk Reduction Strategies – Improved Quality of Health. 1999 ISBN 0-7923-5906-2
14. R. Paoletti, A. Poli and A.S. Jackson (eds.): *Creatine.* From Basic Science to Clinical Application. 2000 ISBN 0-7923-6118-0
15. R. Paoletti, A.L. Klatsky, A. Poli, S. Zakhari (eds.): *Moderate Alcohol Consumption and Cardiovascular Disease.* 2000 ISBN 0-7923-6572-0
16. B. Samuelsson, R. Paoletti, G.C. Folco, E. Granström and S. Nicosia (eds.): *Advances in Prostaglandin and Leukotriene Research: Basic Science and New Clinical Applications* 2001 ISBN 1-4020-0146-0

KLUWER ACADEMIC PUBLISHERS – DORDRECHT / BOSTON / LONDON

GPSR Compliance
The European Union's (EU) General Product Safety Regulation (GPSR) is a set of rules that requires consumer products to be safe and our obligations to ensure this.

If you have any concerns about our products, you can contact us on

ProductSafety@springernature.com

In case Publisher is established outside the EU, the EU authorized representative is:

Springer Nature Customer Service Center GmbH
Europaplatz 3
69115 Heidelberg, Germany

www.ingramcontent.com/pod-product-compliance
Ingram Content Group UK Ltd.
Pitfield, Milton Keynes, MK11 3LW, UK
UKHW020834270526
471509UK00003B/69

* 9 7 8 1 4 0 2 0 0 1 4 6 8 *